Indigenous people and the Pilbara mining boom: A baseline for regional participation

John Taylor and Benedict Scambary

ANU
THE AUSTRALIAN NATIONAL UNIVERSITY

E PRESS

Centre for Aboriginal Economic Policy Research
The Australian National University, Canberra

Research Monograph No. 25
2006

ANU

E PRESS

Published by ANU E Press
The Australian National University
Canberra ACT 0200, Australia
Email: anuepress@anu.edu.au
Web: http://epress.anu.edu.au

National Library of Australia
Cataloguing-in-publication entry.

Taylor, John, 1953 -
Indigenous people and the Pilbara mining boom: A baseline for regional
participation

Bibliography
ISBN 1 9209424 0 8
ISBN 1 9209425 4 8 (Online document)

1. Aboriginal Australians - Western Australia - Pilbara - Economic conditions.
2. Community development - Western Australia - Pilbara. 3. Sustainable
development - Western Australia - Pilbara. 4. Mineral industries - Western
Australia - Pilbara. 5. Pilbara (W.A.) - Economic conditions. I. Scambary, B. II.
Australian National University. Centre for Aboriginal Economic Policy Research.
III. Title. (Series : Research monograph (Australian National University. Centre for
Aboriginal Economic Policy Research) ; no. 25).

362.84991509413

Cover design by Brendon McKinley

Table of Contents

List of Figures

List of Tables

Foreword

This is the latest in what is developing into a series of CAEPR monographs stimulated by the interest of mining companies to better understand the social and economic landscape of regions within which they operate. It stems from an approach made to the Centre for Aboriginal Economic Policy Research (CAEPR) by Jeff Wilkie of Pilbara Iron who sought to establish a detailed profile of the Pilbara population in order to assist company and public policy discussions on Indigenous engagement in the context of rapidly expanding mining activity in the Pilbara region. We acknowledge the financial and logistical support of Pilbara Iron that facilitated the undertaking of the research.

As an output, this monograph sits well within the framework of the Australia Research Council Linkage project, *Indigenous Community Organisations and Miners: Partnering Sustainable Regional Development?* This is a three-year research partnership between CAEPR and industry partners Rio Tinto and the Committee for Economic Development of Australia (CEDA). The project focuses on case studies of the Pilbara region of Western Australia, the Kakadu region of the Northern Territory, and the Gulf region of Queensland where major resource development agreements have been negotiated between the developers and Indigenous people and their organisations.

At the core of this project lies an examination of the conditions that might allow sustainable development for Indigenous people residing in mine hinterlands. The present study addresses this issue by detailing the socio-economic structural constraints on the supply of Indigenous labour in the Pilbara region against a background of increasing labour demand. John Taylor and Benedict Scambary's comprehensive profile provides an invaluable baseline against which the future outcomes from mining agreements and regional initiatives might be gauged. A soon-to-be completed complementary project monograph to be published in 2006 will focus on the considerable diversity in the values and aspirations of the Indigenous people of mine hinterlands, and the fact that these are not limited to involvement in mining-related employment or business development, but may focus on other, more specifically Indigenous 'social or community economies'.

The capacity of resource developers to realise employment and other goals, as often set out in agreements with Indigenous people, is a matter of considerable interest when considering the economic development prospects of Indigenous people in mine hinterlands. I commend Pilbara Iron for facilitating a flow of information that will ensure greater discussion of such issues.

Professor Jon Altman
Director, CAEPR
December 2005

Acknowledgements

We are grateful for the impetus, and financial and logistical support for this study provided by Jeff Wilkie and Bruce Larson of Pilbara Iron. Together with their colleagues at ATAL in Dampier (Mark Simpson in particular), they ensured that no stone was left unturned in the pursuit of essential data. Others who assisted in this process are too numerous to list by department but included Jim Codde, Vivien Gee, Peter Somerford, John Harris, Judith Uren, Luke Drozdowski, Mark Jessop, Robin Wood, Roger Holding, Tom Mulholland, Stephen Smythe, Bob Hay, and Mark Hewitt from key Western Australian government agencies, along with Julie Kirkby of Centrelink and Mohammed Shahidullah and Maria Meere from the ABS. Other invaluable assistance was provided by Ian Satchwell of ACIL Tasman, John Fernandez at the University of Western Australia, Tony Meagher at Monash University, Anne Larson of the Combined Universities Centre for Rural Health in Geraldton, Louis Warren at BHP Billiton, and Brian Hughey and Charmain Tullock from Ngarda Civil and Mining. Simon Hawkins and Adrian Murphy of the Yamatji Marlpa Barna Baba Maaja Aboriginal Corporation were also helpful in their support. At the ANU, Yohannes Kinfu assisted in the preparation of population projections, while Kay Dancey drafted the maps. Finally, we are grateful for the constructive comments on initial drafts from two referees as well as many CAEPR colleagues (notably David Martin). Geoff Buchanan of CAEPR meticulously weeded out any errors, and Frances Morphy spent long hours copy editing the manuscript for publication. Brendon McKinley of ANU E Press was responsible for the layout and design

A note on spellings of Aboriginal words

Aboriginal words in this document fall into the categories of language group names, names of organisations, place names, and native title claim names. There is considerable inconsistency in the orthographies used in these different categories. Examples include the language group name Yinhawanga and the Innawonga Aboriginal Corporation, and the language group Kurrama and the Eastern Guruma Agreement.

Accordingly, the spellings of organisation names are those used by the organisations themselves. Native Title claim names are reproduced as they are on the National Native Title Tribunal's register of native title claims. Where language groups are referred to generally, spellings are consistent with the Wangka Maya Pilbara Aboriginal Language Centre website <http://www.wangkamaya.org.au/languages.htm>, and the AIATSIS ASEDA online catalogue <http://coombs.anu.edu.au/SpecialProj/ASEDA/ >. Place name spellings are consistent with those specified by the Geographic Names Committee of the Western Australian Department of Land Information.

Where Aboriginal words and names are reproduced from secondary sources the spellings have not been corrected in accordance with the above, but left as they originally appeared.

Acronyms and abbreviations

ABS Australian Bureau of Statistics

AIATSIS Australian Institute of Aboriginal and Torres Strait Islander Studies

AIGC Australian Indigenous Geographic Classification

ANU The Australian National University

ANZSIC Australia and New Zealand Standard Industrial Classification

ASCO Australian Standard Classification of Occupations

ASEDA Aboriginal Studies Electronic Data Archive

ASFR age specific fertility rate

ASGC Australian Standard Geographic Classifiaction

ATAL Aboriginal Training and Liaison (Unit, of Pilbara Iron)

ATSIC Aboriginal and Torres Strait Islander Commission

CAEPR Centre for Aboriginal Economic Policy Research

CBO Community Based Order

CDEP Community Development Employment Project(s)

CEDA Committee for the Economic Development of Australia

CHINS Community Housing and Infrastructure Needs Survey

CHIP Community Housing and Infrastructure Program

CHIPS Children's Court and Petty Sessions

CI confidence interval

CRC Crime Research Centre

DHW Department of History and Works

DIDO drive-in-drive-out

DSC Disability Services Commission (WA)

EHNS Environmental Health Needs Survey

ERP estimated resident population

FIFO fly-in-fly-out

FTB Family Tax Benefit

GEHA Government Employees Housing Authority

GIS Global Information System

GMY Gobawarrah Minduarra Yinhawanga

HIPP Health Infrastructure Priority Projects

IA Indigenous Area

IBN Innawonga Banyjima Nyiyaparli

ICCP Indigenous Communities Coordination Project

IC9 Ninth Revision, International Classification of Diseases

IL Indigenous Location

ISO Intensive Supervision Order

KRSIS Kakadu Social Impact Study

NAHS National Aboriginal Health Strategy

NATSIS National Aboriginal and Torres Strait Islander Survey

NATSISS National Aboriginal and Torres Strait Islander Social Survey

NILF not in the labour force

PNTS Pilbara Native Title Service

PoE Place of enumeration

PRR Partial paternity rate

PDM population density measure

P49 Police Offence Information System

SCRGSP Steering Committee for the Review of Government Service Provision

SD Statistical Division

SLA Statistical Local Area

SRR standardised mortality rate ratio

STEP Structured Training and Employment Program

TAFE Technical and Further Education

TES Total Employment Services

TFR total fertility rate

TOMS Total Offender Management System

VET vocational education and training

WDO Work and Development Order

WHO World Health Organisation

WMPALC Wangka Maya Pilbara Aboriginal Language Centre

1. Profiling outcomes

In developing the ARC linkage project *Indigenous Community Organisations and Miners: Partnering Sustainable Regional Development* between the Centre for Aboriginal Economic Policy Research (CAEPR) at the Australian National University (ANU) and Rio Tinto, it was noted that the number of agreements between mining companies and Indigenous community or regional organisations had grown substantially over the past two decades. It was also noted that a degree of optimism prevailed in the early 1980s that agreements such as these, many with significant financial benefit packages, would make a difference to the marginal economic situation of Indigenous beneficiaries. However, research to date indicates that, for a complex set of reasons, Indigenous economic status has changed little in recent decades – dependence on government remains high and the relative economic status of Indigenous people residing adjacent to major long-life mines is similar to that of Indigenous people elsewhere in regional and remote Australia. This unexpected outcome was clearly demonstrated in the Kakadu Region Social Impact Study (KRSIS) (Taylor 1999), and more recently in regional profiling work associated with the Argyle Diamond Mine Participation Agreement (Taylor 2004a).

According to Altman (2001), this situation of stasis partly reflects the limited capacity of Indigenous community organisations both to cope with the impacts of, and take advantage of, large-scale operations. On the other hand, it is also seen that such organisations and the people they represent may have ambivalent responses to the potential cultural assimilation implied by their increasing integration into a market economy and its monetisation of many aspects of social life. A third key factor proposed is the attitudes and responses of mining companies and governments, and their inability to comprehend the extent of historic Aboriginal disadvantage and strain on the social fabric of societies so radically affected by colonisation.

Partly in response to these issues, there has been a concerted effort in recent years by some major mining companies to address aspects of this regional development problem. Of particular relevance to the present study is a growing recognition that sound community relations and the pursuit of sustainable regional economies provide the necessary foundations for a social licence to operate – a factor of rising market significance (Harvey 2002; Trebeck 2003). In establishing this licence, the construction of statistical baseline profiles of social and economic conditions within mine hinterlands (with a focus on establishing the relative situation of Indigenous populations) is regarded as a crucial input by significant players in the mining industry (Harvey 2001; Harvey & Brereton 2005). Accordingly, Indigenous representative bodies (such as Land Councils) have shown interest in supporting such activity as in the case of the KRSIS

(Taylor 1999) and the Argyle Diamond Mine Participation Agreement (Taylor 2004a). Such profiles assist in the formulation and subsequent monitoring of company and Indigenous stakeholder actions designed to increase Indigenous participation in regional economies. Specifically, they help to establish the range and quantum of needs for regional planning, to identify opportunities and constraints for enhanced participation, and to assess the effectiveness of actions undertaken.

The timing of this study, then, is deliberate for two reasons. First, it seeks to respond to the call for detailed profiles of regional populations set within a framework of understanding the dynamics of Indigenous labour demand and supply. Second, it comes at the outset (or somewhat into the commencement) of potentially the largest escalation in mining activity in Australian history that has emerged primarily as a consequence of the successful procurement of export contracts (mostly to China) by Pilbara-based transnational resource companies. Given the scale of planned new mineral output, new infrastructure development, and associated regional multipliers, major social and economic impacts on Indigenous communities in the Pilbara region are to be expected. Admittedly, anticipated developments do not always eventuate, not least in this particular region (Linge 1980), but such is the scale of committed investment and associated infrastructure that even sub-optimal outcomes will make their mark. In order to anticipate, plan for, and subsequently assess the impact of these developments, it is timely that a profile of existing socio-economic conditions in the region be undertaken. Accordingly, this study develops and presents social indicators for the Indigenous and non-Indigenous populations of the Pilbara region. Its aim is to assist in the implementation and subsequent monitoring of company and Indigenous stakeholder activities aimed at meeting particular goals in terms of Indigenous participation in the regional economy.

To this extent, the construction of a statistical profile falls analytically within the realm of Social Impact Assessment, this being an area of systematic inquiry which seeks to investigate and understand the social and economic consequences of planned change and the processes involved in that change (Ross 1990). Analysis of this type has an established history in Western Australia, most notably in the East Kimberley where (as the East Kimberley Impact Assessment Project under the direction of the late H.C. Coombs) it emerged as an essential feature of the public policy response to the initial development of Argyle Diamond Mine (Coombs et al. 1989; Dillon 1990). Surprisingly, given the much larger scale and greater longevity of mining activity in the Pilbara, no comparable assessment work has been conducted in this region.

Whilst it is fair to say that the East Kimberley work occurred largely at the insistence of local Indigenous communities (Dillon 1990), it is equally true that the need for monitoring of Indigenous and regional social and economic conditions is now enshrined in Rio Tinto policy governing relations with local communities (Harvey 2002), and is implicit in the forward planning strategies of companies such as Pilbara Iron as expressed in its Indigenous Employment Strategy (Aboriginal Training and Liaison unit (ATAL) 2005). Likewise, the content of Indigenous Land Use Agreements across the Pilbara between mining companies and traditional owners invariably include stipulation of opportunities for employment, training, education, and enterprise development. Accordingly, the current political economy of mining in the Pilbara demands that Indigenous communities are positioned to more fully avail themselves of economic opportunities as they emerge (Harvey 2002). The strengthening of activities by Hamersley Iron's (now Pilbara Iron's) ATAL unit is one important manifestation of this structural change that has relevance to the current exercise.

In line with these developments, the focus of social scientific research on Indigenous peoples and the mining industry has shifted somewhat in recent years away from solely impact assessment towards supporting a social sustainability/regional development agenda that was first mooted by Cousins and Nieuwenhuysen (1984) in their pioneering and wide-ranging analysis of the position of Indigenous people in relation to the Australian mining industry. This paradigm shift is in line with recent corporate restructuring aimed at bolstering the social licence to operate, and it also reflects enhanced Indigenous stakeholder interest in securing increased participation and sustainable benefits (Trebeck 2003). Thus, the task of profiling is now more applied than before with greater focus on establishing the range and quantum of needs for regional planning, and identifying the opportunities and constraints for enhanced participation. For example, a common fallacy is that Indigenous labour force participation in remote areas is low because remote areas have no labour market. The fact is, a substantial mainstream labour market does exist in remote Australia (not least in the mining industry), and the reasons for low Indigenous participation are complex (Taylor 2005a). In attempting to understand this disparity between labour demand and supply, it is necessary to quantify basic components of the regional economy such as those already in work, those who might be employed, and those who (for a variety of reasons) are unlikely to acquire mainstream employment.

The framework for such an exercise in the Pilbara has been the emergence of significant land use (mining) agreements between Rio Tinto and Indigenous stakeholders, within which there is an increasing recognition that realisation of the benefits of mining to local populations, both in the production phase and beyond, requires the development of a sustainable mixed regional economy. This, in turn, necessitates the inclusion of an enhanced Indigenous capacity to engage and participate in the regional economy either through direct mainstream employment or more traditional pursuits. Such intent necessarily widens the scope of any impact analysis beyond the relatively narrow geographic focus of individual mining agreements, or company areas of impact, to encompass a more functional definition of 'area affected' based on some measure of regionally integrated social, economic and administrative interaction. The appropriate scale at which this integration is captured for the present study is the Pilbara Statistical Division (SD) incorporating the four Shires of Ashburton, East Pilbara, Roebourne, and Port Hedland (Fig. 1.1). This more or less represents the northern jurisdiction of the Pilbara Native Title Service (PNTS) which is the Native Title Representative Body for the Pilbara under the umbrella of the Yamatji Marlpa Barna Baba Maaja Aboriginal Corporation (Yamatji Land and Sea Council). It also defines the spatial extent of a population from which any future Indigenous regional labour force is to be mostly drawn.

Figure 1.1. Statistical geography of the Pilbara region

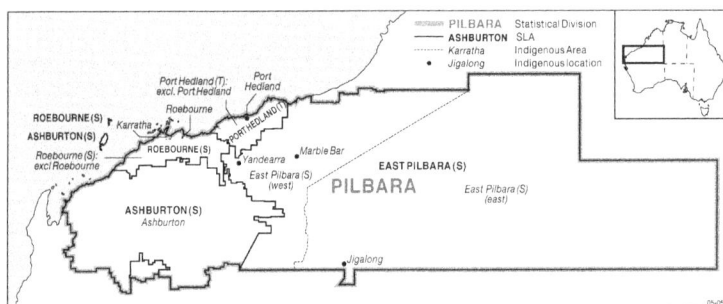

Methods

Two main tasks are assumed under the present exercise. The first is a monitoring role enabled by the provision of a baseline profile of existing conditions against which an assessment of past and future change can be made. The second is a predictive role, or at least an anticipation of the possible effects of proposed developments. Both of these require statistical profiling, which in turn involves analysis and measurement of the social and economic conditions of the population as defined. While these tasks might seem clear enough, the manner in which they have been carried out in particular cases has varied (Coombs et al. 1989; Kakadu Social Impact Study (KRSIS) 1997; Kesteven 1986; Taylor 1999, 2004a; Taylor, Bern & Senior 2000). In the present study, the aim is to establish the relative socioeconomic status of Indigenous people in the Pilbara and to consider the prospects of likely impact on this over the establishment period of current mining expansion (roughly to 2016).

In constructing this statistical profile, a range of data are compiled from a variety of published and unpublished sources including the Census of Population and Housing, and administrative data sets held by Commonwealth and Western Australian government departments, Pilbara Iron, and other regionally-based institutions. Because of the specific focus on generating statistical information, reference to literature that describes aspects of social and economic life in the region, both past and present, is limited to instances where this provides a key source of statistical data or assists in its interpretation.

The scope of the profile is limited to several key areas that form the basis of policy interest and potential intervention. These include the demographic structure and residence patterns of the regional population, labour force status, education and training, income, welfare, housing, justice and health status. For each of these categories, the aim is to identify and describe the main characteristics of the population and to highlight outstanding features in the data. As far as possible, time series are also compiled to establish the trajectory of recent socioeconomic change. Also, where appropriate, comment is made on the adequacy of coverage and robustness of available data, while comparison is drawn with non-Indigenous people in the region as well as with Indigenous people elsewhere in Western Australia.

In profiling the circumstances of Indigenous and non-Indigenous Australians, analysts rely heavily on census data for many key indicators. This has a number of advantages given the comprehensive scope of coverage and the application of standard measures. However, there are drawbacks too, especially for the Indigenous population. First of all, an over-reliance on five-yearly census data means that information on Indigenous populations required, say in 2006, is five years old. There is also a problem of coverage, both in terms of population numbers and population characteristics. Overall in Australia, the undercount of Indigenous peoples is estimated to be around 6 to 7 per cent, although this varies geographically. Unfortunately, no direct estimate of this variation is established in sparsely settled areas. Non-response to census questions also occurs, with relatively high rates observed for many Indigenous population characteristics. While little can be done about this loss of information, the ABS does establish post-censal estimates of the Indigenous population in an attempt to adjust for undercount and non-response to the Indigenous status question, and herein lies a solution to the first problem of coverage raised above.

The approach taken here is to view the census as a very large sample survey, with the key output being population rates rather than population levels. Rates established net of non-response (on the assumption that the latter are evenly distributed for each population characteristic) can then be applied to population estimates – initially to the estimate for the census year, and then to population projections from the census year on the assumption that the observed rates remain constant. While this assumption of constancy might be seen as unrealistic, it should be noted that one of the (unfortunate) features of many Indigenous social indicators in Western Australia (and more generally) over the past two decades (such as labour force status, income, education, and housing) has been their relative stability (Hunter 2004; Jones 1994; National Centre for Social

Applications of GIS 2003; Taylor 1997; Taylor & Roach 1994). It is also true that social indicator rates, by their very nature, are unlikely to drastically alter over short periods of time requiring, as they do, a substantial shift in levels in order to effect change. This is especially so among rapidly growing populations, such as that found in the Pilbara.

Perhaps more telling, from the point of view of data quality, are concerns about the capacity of census (and other) data to provide a meaningful representation of the social and economic status of Indigenous people in the region. With census data, for example, there are concerns about the cultural relevance of information obtained from an instrument principally designed to establish the characteristics of mainstream Australian life (Smith 1991; Morphy 2002). Thus, having observed the 2001 Census count first hand at a Northern Territory outstation, Morphy (2002) has described the process of enumeration as a 'collision of systems' and concludes that census questions lack cross-cultural fit and can produce answers that are often close to nonsensical. Equally, while social indicators report on observable population characteristics, they reveal nothing about more behavioural population attributes such as individual and community priorities and aspirations. In short, when set against mainstream outcomes they describe the relative condition of Indigenous people, but contain no Indigenous voice.

Accordingly, this form of statistical profiling using mainstream statistical data sources can be seen as a means of 'rapid appraisal' and rightly criticised as lacking community input thereby restricting its relevance and representativeness, certainly from a cross-cultural perspective (Birckhead 1999; Walsh & Mitchell 2002). Thus, as an adjunct to the compilation of statistical information, the current exercise incorporates qualitative data gathered by semi-structured discussions with randomly selected Indigenous informants. The aim here was to elicit perspectives on the types of opportunities and constraints that local Indigenous people consider significant in terms of their participation in the regional economy. It was initially proposed to undertake these interviews in a number of locations across the Pilbara. However, given the time constraints of just 10 days fieldwork, and the vastness of the study region, a total of just 25 interviews were conducted mostly within the central Pilbara (as well as some coastal centres) close to where the bulk of Pilbara Iron's operations are located.

Despite this geographic focus, interviewees were not confined to land owning groups associated with the operations of Pilbara Iron. Under the umbrella of the CAEPR–Rio Tinto Australia Research Council linkage project, the majority of those interviewed were drawn from the Banyjima, Yinhawangka, and Nyiyaparli language groups who are associated with the Yandi Land Use Agreement. Initially it was also hoped to interview people from a range of age, gender, and socio-economic groups. However, it proved difficult to engage with youth as those in the 18-25 year age group who were approached were unengaged by the research questions. The style of interview was an open discussion, often conducted collectively with more than one person, with the focus built around issues arising from the main statistical profile of regional Indigenous socio-economic status and opportunities for improvement. In summary, the trends highlighted were a general increase in the Indigenous population of the Pilbara, an increase in non-Indigenous population due to industrial expansion in the region, poor educational and health outcomes for Indigenous people, overcrowding of houses occupied by Indigenous people, poor labour force representation, and consequent low incomes. Discussions then focused on people's own experience of these factors, and their ideas and thoughts regarding their cause and possible resolution. In addition people were asked to outline their aspirations for the future in relation to the preliminary findings of the demographic study. In line with ANU ethics guidelines, permissions for these discussions were sought, and the identity of interviewees and any identifiable characteristics are suppressed.

Obviously, the aim of these narratives was not to present a totality of Indigenous views of life and futures in the Pilbara, not least because more comprehensive and wide-ranging compendiums already exist (Olive 1997; Wangka Maya Pilbara Aboriginal Language Centre (WMPALC) 2001). Rather, it was simply to provide a human frame of reference for the more statistical elements of the regional profile. By means of cross-reference throught the text, this blending of quantitative and qualitative data allows some appreciation of the real-life circumstances that lie behind and contribute to the raw statistics.

Statistical geography

It is fortunate that the Pilbara and its constituent sub-regions form a relatively stable geography over time within the Australian Standard Geographic Classification (ASGC) of the Australian Bureau of Statistics (ABS). This provides a basis for some consideration of change in socioeconomic conditions since the modern era of mining in the Pilbara region commenced in the 1960s. As for current indicators, the focus is mostly on the Pilbara SD, given the broad regional remit of the study, whilst the four Statistical Local Areas (SLAs) of Ashburton, East Pilbara, Roebourne and Port Hedland provide the main platform for sub-regional analysis. More detailed geography is provided by the Australian Indigenous Geographic Classification (AIGC) that includes 10 Indigenous Areas (IAs) and 23 Indigenous Locations (ILs) within the Pilbara SD. These various statistical boundaries are shown for reference purposes in Fig. 1.1. Further information at even greater levels of spatial disaggregation is also available from the 2001 Community Housing and Infrastructure Needs Survey (CHINS) that reports at the level of 33 discrete communities within the Pilbara.[1]

The adoption of this geographic frame is not to deny that the social reality, for both Indigenous (and non-Indigenous) people in the region, is one of social, cultural, and economic interconnectedness between this region and elsewhere in Western Australia (especially Perth in the case of the non-Indigenous population). One manifestation of this is the frequent movement of individuals, groups and families into and out of the Pilbara, rendering an unambiguous definition of the regional population problematic. A further constraint imposed by this geography is on any representation of socio-economic conditions according to cultural boundaries – for example, as presented by the configuration of Indigenous-owned lands and native title claim areas - or company geography – for example, as presented by the distribution of particular mine sites and infrastructure networks.

[1] Discrete communities are defined by the ABS as geographic locations that are bounded by physical or cadastral boundaries, and inhabited or intended to be inhabited predominantly by Indigenous people (more than 50 per cent), with housing and infrastructure that is either owned or managed on a community basis (ABS 2002a).

2. Demography of the Pilbara region

A range of counts and estimates are available for the Indigenous and non-Indigenous populations of the Pilbara and its constituent parts. For example, the ABS provides a de facto count of people who were deemed to be present in the region on each census night (7 August 2001 at the most recent census). Then there is a de jure count of people across Australia who indicate that the Pilbara is their usual place of residence on census night. These two counts are also available for SLAs and IAs found within the region, while de facto counts are available for select ILs. It is also worth noting that the ABS gathers information on estimated population numbers for all discrete Indigenous communities, no matter how small, via the CHINS which is now rolled out three months ahead of each census. Finally, in recognition of the fact that the census fails to count some people, the ABS develops post-censal estimates of the 'true' resident population by augmenting their SLA usual residence counts according to an estimate of those missed (undercount), along with other demographic adjustments. This produces an Estimated Resident Population (ERP), which, in effect, becomes the official population of each SLA for the purposes of electoral representation and financial distributions.

It should be emphasised that official ERPs are only available at the SLA-level, and so estimates of the population within the region at levels below this have to be derived by ratio allocation of the overall Pilbara ERP to constituent parts. In terms of the present exercise that is concerned with the relative status of Indigenous people, it is helpful that separate calculations of Indigenous and non-Indigenous ERPs have also been made by the ABS since 1996. These are constructed by applying differential estimated undercount rates, and by distributing (pro rata) those usual residents who did not answer the ethnicity question on the census form.

As to the measurement of demographic change in the region since mining operations commenced more than 40 years ago, it is also fortunate that the ABS has retained the statistical geography of the Pilbara SD over this period. This provides for a fairly complete analysis of population change over time, including a measure of variable growth and redistribution within the Pilbara. Of course, many of these changes reflect the varying fortunes of the mining industry over the past few decades, while for Indigenous people they reflect the myriad influences on their location due to mining development, changes in the pastoral industry and in government policy, and in their own access to kin and country.

Population size

At the 2001 Census, a total of 42 411 persons were counted by the ABS as present on census night (7 August 2001) in the Pilbara (Table 2.1 p. 10). Of these, 5736 indicated an Indigenous status in response to the census question on ethnicity, and 33 622 indicated non-Indigenous status. Thus, as many as 3053 individuals (7%) provided no response to this question, and so their Indigenous status was indeterminate. Of the entire population counted within Australia on census night, a smaller total (37 135) nominated one of the Pilbara SLAs as their usual place of residence. Thus, the overall usual residence count was 12 per cent lower than the place of enumeration count, with as many as 5276 individuals counted in the region as temporary residents from elsewhere.

Table 2.1. Indigenous and non-Indigenous census counts and post-censal estimates: Pilbara SD,[a] 2001

	Indigenous	Non-Indigenous	Not stated	Total
Census count (de facto)	5736	33 622	3053	42 411
Usual residence count (de jure)	5579	28 744	2812	37 135
Estimated usual residents (ERP)	6514	32 947	n/a	39 463

[a]Incorporates Ashburton, Roebourne, Port Hedland, and East Pilbara SLAs.
Source: ABS customised tables.

It is clear that almost all of these temporary people are non-Indigenous since the number of Indigenous usual residents of the Pilbara was almost identical to the number counted as present there on census night. On this evidence, more than 10 per cent of the non-Indigenous population present in the Pilbara at any given time is visiting the region from somewhere else in Australia, at least during the dry season when the census is conducted. This issue will be revisited when analysing the composition of the Pilbara labour force; suffice it to say here that temporary residents form an important part of the Pilbara demography and add complexity to the task of estimating and projecting population levels.

ERP figures for the Pilbara are shown in the final row of Table 2.1. As noted earlier, these purport to represent 'true' levels of the Indigenous and non-Indigenous resident populations of the region. However, when interpreting these, it is important to note that ABS ERPs have been observed to differ from other (unofficial) population estimates generated by alternate means (Taylor & Bell 2001, 2003). Also for noting are methodological tendencies within the special procedures adopted by the ABS in remote communities and urban town camps in northern Australia that are likely to produce an undercount of Indigenous people (Martin & Taylor 1996; Sanders 2002; Taylor 2005b). This places an onus on the standard ERP methodology to adequately compensate for these shortcomings, and the capacity to achieve this has been questioned especially given the lack of a post-enumeration survey check of enumeration coverage in remote areas (Taylor & Bell 2003). However, Table 2.1 does show that the ERP upward adjustment to the Indigenous usual residence count was quite substantial at 16.8 per cent – anything higher than this would require compelling corroborative evidence, and none is readily apparent.

The figures in Table 2.1 raise a subtle but important point about regional population shares and quotas. Strictly speaking, the correct means for deriving the Indigenous share of the Pilbara regional population is to use the ERP figures. On this basis, Indigenous people represent 16.5 per cent of the Pilbara resident population. However, as we have seen, there are many others (over 5000) who are likely to be found in the Pilbara at any given time (at least on the basis of the 2001 reported levels), and the suspicion is that many of these are workers or are at least potentially available to compete for work. While this point will be further developed later, there might also be a case for arguing that the Indigenous share should be derived using the de facto population as the denominator, in which case in 2001 the share would have been 14.5 per cent (excluding those who did not state Indigenous status). Admittedly, this produces only minor variation and so the point may seem trivial. However, in a scenario where fly-in-fly-out (FIFO) regimes might be greatly extended, and where large temporary workforces can move into the region during ramp-up phases of resource development, the de facto population can become very substantial. If, in such instances, ratio shares are based on de facto rather than de jure counts then the Indigenous share could be considerably diminished. We should also note the importance of accounting for relative age distribution (since the Indigenous regional share of working-age population is much lower than the share of total population), while alternate projections of future population can also produce quite different outcomes in terms of rising and falling shares.

Sub-regions

The primary sub-regional breakdown of the Pilbara is by the four Shires that correspond to SLAs. Table 2.2 p. 12 shows the distribution of Indigenous and non-Indigenous ERP populations in each of these. The coast versus inland divide in the distribution of the Pilbara population is readily evident, with Roebourne and Port Hedland Shires standing out with the largest populations and accounting for 70 per cent of overall numbers due to the presence of major port, processing, and service towns. The less populous inland shires of Ashburton and East Pilbara reflect the location of various mine sites, mining towns, and discrete Indigenous communities. Within this broad framework variation does occur between Indigenous and non-Indigenous distributions. Thus, Indigenous people are more heavily represented in East Pilbara Shire, and considerably outweighed in Roebourne Shire. Of course, these numbers reflect ERPs only. If we were to add to this the presence of temporary residents (as estimated from the place of enumeration counts) then the numbers in the non-Indigenous column would rise overall by almost 3000, with the distribution of these temporary numbers fairly evenly distributed across the Shires.

Table 2.2. Indigenous and non-Indigenous population[a] distribution by Pilbara SLAs, 2001

SLA	Indigenous	Non-Indigenous	Total
Ashburton	732 (11.2)[b]	5213 (15.8)	5945 (15.0)
Roebourne	1878 (28.8)	13 180 (40.0)	15 058 (38.1)
East Pilbara	1611 (24.7)	4232 (12.8)	5843 (14.8)
Port Hedland	2293 (35.2)	10 322 (31.3)	12 615 (32.0)
Total	6514 (100.0)	32 947 (100.0)	39 461 (100.0)

[a]ERPs

[b]Percentages in parentheses.

As noted, ERP figures are available only at SLA-level. However, one way to develop similar estimates of the regional population according to a geography that more closely relates to the distribution of Indigenous population in different parts of the region, is to divide up the Indigenous and non-Indigenous ERPs according to the observed pro rata share of each IA. Results of this ratio allocation are shown in Table 2.3 p. 12.

Table 2.3. Ratio allocation of 2001 Pilbara ERPs to Indigenous Areas in the Pilbara SD

Indigenous Area	Indigenous % of Pilbara UR[a] count	Non-Indigenous % of Pilbara UR resident	Derived Indigenous ERP	Derived Non-Indigenous ERP	Derived total ERP	Indigenous % of total derived ERP
Roebourne	10.2	0.6	663	205	869	76.4
Roebourne bal	8.5	9.6	556	3159	3715	15.0
Karratha	10.2	31.2	666	10 290	10 955	6.1
Ashburton	11.0	16.0	718	5273	5991	12.0
East Pilbara (W)	3.7	1.6	238	517	755	31.5
East Pilbara (E)	11.5	10.8	750	3565	4315	17.4
Jigalong	4.9	0.1	321	18	340	94.6
Marble Bar	2.1	0.3	134	83	217	61.9
Yandeyarra	3.4	0.0	220	13	232	94.6
Port Hedland (T)	2.6	0.2	167	81	248	67.2
Port Hedland	32.0	29.6	2082	9744	11 826	17.6
Total	100	100	6514	32 947	39 463	16.5

[a]UR = usual residence.

Thus, the Indigenous usual residence count in Karratha (755) represented 10.2 per cent of the Indigenous usual residence count for the whole of the Pilbara. This same percentage of the Pilbara Indigenous ERP produces an Indigenous population estimate for Karratha of 666. In turn, the equivalent non-Indigenous proportion is 31.2 per cent, which produces an estimate of 10 290 for the non-Indigenous population of Karratha. All told, then, the 2001 ERP of Karratha is calibrated at almost 11 000, only 6 per cent of which is Indigenous.

A substantial proportion of both the Indigenous and non-Indigenous populations of the Pilbara is located in Port Hedland. At the same time, many other areas of the Pilbara (Roebourne, Jigalong, Marble Bar, Yandeyarra, and the areas adjacent to Port Hedland) have overwhelmingly Indigenous populations, while in some other areas (East Pilbara W) the Indigenous population share is also way above the regional average. The main exceptions are Karratha and Ashburton where company and other towns predominate. The simple point here is that over vast tracts of the Pilbara region, the 16 per cent global Indigenous share statistic can be misleading as large parts of the country away from the demographic influence of urban centres and mine sites remain essentially Indigenous domains where Indigenous people and their institutions predominate.

Population growth

Time series analysis of these estimated populations is rendered problematic by the lack of official Indigenous/non-Indigenous estimates at the SLA-level prior to 1996, plus the fact that intercensal estimates prepared by the ABS do not include an Indigenous component. Nonetheless, total estimates are available for the Pilbara back to 1976. For these earlier dates, respective census counts can be used to derive an Indigenous share of population with which to pro rata the total ERPs to Indigenous and non-Indigenous categories. This is crude, but effective, in establishing relative growth trends. The results of this manipulation are shown in Table 2.4 p. 13, while the same data are shown graphically in Figure 2.1 p. 14.

Table 2.4. ERP by Indigenous status: Pilbara SD, 1976–2006

Year	Indigenous[a]	Non-Indigenous	Total
1976	n/a	n/a	38 750
1981	4336	42 294	46 630
1986	4582	43 147	47 729
1991	5400	41 150	46 550
1996	5721	34 705	40 426
2001	6514	32 947	39 461
2006[a]	7141	35 759	42 900

[a]Indigenous estimate based on projection by Yohannes Kinfu and authors. The estimate of total population in 2006 is derived from the Western Australia Department for Infrastructure and Planning. All non-Indigenous estimates are derived as a residual.

While the Indigenous population has grown steadily over the past couple of decades, and continues to do so, the non-Indigenous population (being an essentially migrant-based group) has waxed, then waned, and waxed again in response to the cycles of regional economic fortune. Thus, after rising to a peak in the mid-1980s, the non-Indigenous population declined for more than a decade, and has revived again in recent years. Over the same period, the Indigenous population has steadily grown. Consequently, from being just 9 per cent of the Pilbara resident population in 1981, it is estimated that Indigenous people will account for 16.6 per cent of the regional total by 2006. As can be seen, it is actually growth in the Indigenous population that has provided a brake on regional population decline for much of the past two decades.

Figure 2.1. Indigenous and non-Indigenous estimated population levels: Pilbara SD, 1981–2006

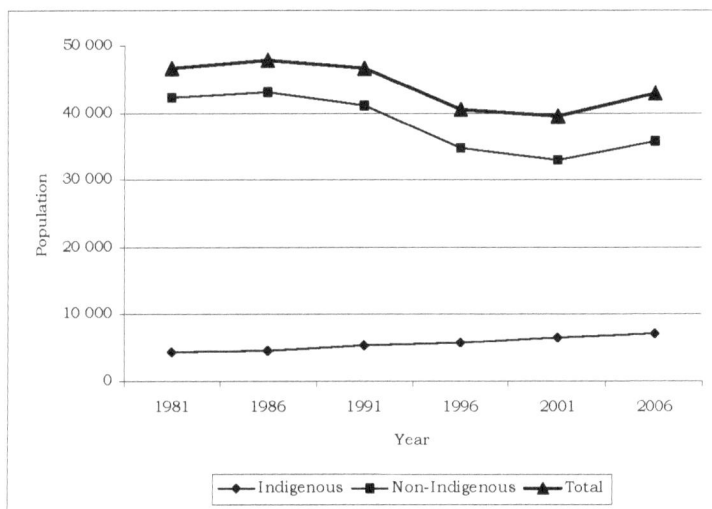

Source: ABS; Western Australia Department for Planning and Infrastructure; Indigenous population projections from calculations by Dr Yohannes Kinfu and author.

Population distribution

The nature and extent of Indigenous participation in the regional economy can be greatly affected by the spatial distribution and residential structure of the population. One way of depicting this distribution has already been presented using Indigenous Areas (Table 2.3). While this revealed some concentration of Indigenous population in urban areas, the full extent of this was hidden.

According to 2001 Census data, fully 67 per cent of the Pilbara's Indigenous population is resident in one of the 10 main towns including (in rank order) Port Hedland, Karratha, Roebourne, Wickham, Tom Price, Newman, Onslow, Dampier, Paraburdoo, and Pannawonica. While there has undoubtedly been a shift over time towards more urban residence, there is a problem in reading too much into this trend as the particular census geographic units employed here mask considerable diversity of residential circumstances of Indigenous people within the region.

Fortunately, the ABS has acquired a new means of representing Indigenous population distribution via the CHINS. This is rolled out ahead of the national census and includes an estimate of the resident population of all discrete Indigenous communities across the nation. According to the 2001 CHINS, a total of 33 discrete Indigenous communities were located within the Pilbara region with a collective estimated population of 2246.[1] A number of these communities are close to urban centres, while most are remote small townships and outstations. Fig. 2.2 shows the size of these settlements in rank size order.

Aside from a couple of localities that have populations of around 300 persons (Jigalong and Yandeyarra), all others are smaller with many having less than 50 persons. These localities are widely scattered and represent living areas close to country (Fig. 2.3). What they signify are individual and collective choices to pursue non-urban lifestyles more in tune with customary norms. Ironically, given the location of many of these settlements, residents of most discrete communities find themselves physically much closer to inland mine sites than their more urbanised Indigenous counterparts along the coast. Notwithstanding this, factors other than relative proximity play a role in determining participation in the regional economy.

Figure 2.2. Rank size distribution of discrete Indigenous communities in the Pilbara SD, 2001

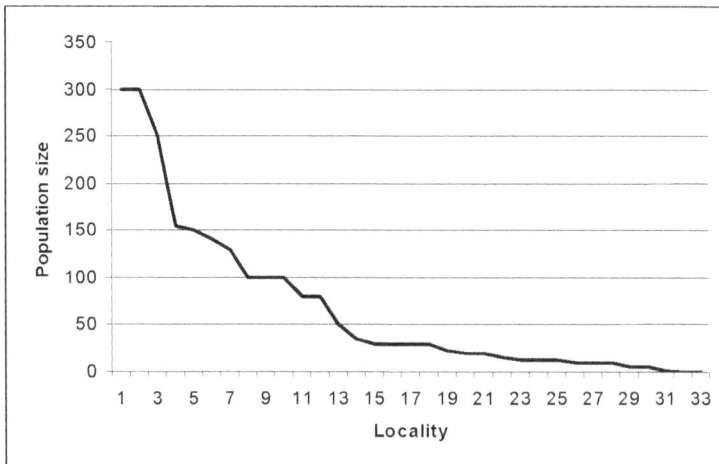

Source: ABS CHINS 2001 Confidentialised Unit Record File.

---

[1]This CHINS reports estimates of the usual resident population of each community based on information provided to survey collectors by key informants in community housing organisations, councils and resource centres. The figures are more akin to service populations.

Figure 2.3. Distribution of discrete Indigenous communities in the Pilbara SD, 2001

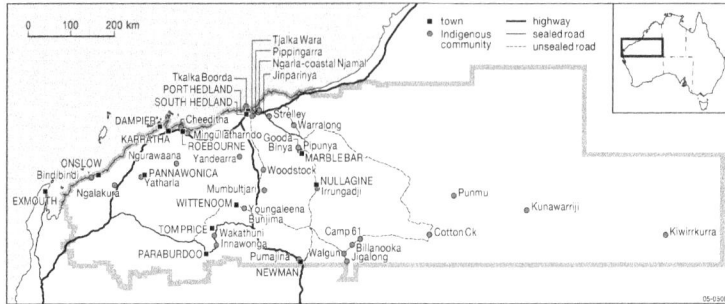

Source: ABS CHINS 2001 Confidentialised Unit Record File.

Age composition

A further demographic feature that has implications for current and future economic status is the contrast between the age distribution of the Indigenous and non-Indigenous populations as shown in Fig. 2.4 for the Pilbara as a whole. For the Indigenous population, several features are noteworthy. First, the relatively broad, though progressively narrowing, base of the age pyramid describes a population with recently reduced, but still relatively high fertility (a total fertility rate of 2.9). Second, the rapid taper with advancing age highlights continued high adult mortality. Third, the relative absence of young adults aged 15–24 suggests out-migration at these ages for both males and females. Fourth, uniformity in the decline of population with age suggests net inter-regional migration balance. Finally, relatively large numbers of women in the child-bearing ages, and even larger cohorts beneath them, indicates high potential for future growth in numbers.

Figure 2.4. Distribution of the Indigenous and non-Indigenous populations[2] of the Pilbara SD by age and sex, 2001

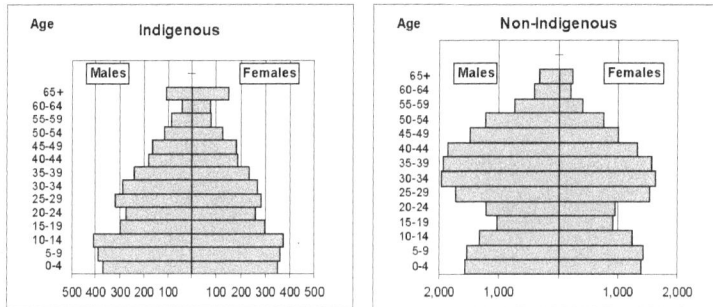

[2]Based on 2001 ABS ERP.

By contrast, the non-Indigenous age distribution is typical of a population that is subject to selective inter-regional migration producing net gains among those of working age and their accompanying children, and net losses in the teen-ages and at older working and retirement ages. Also evident is a predominance of male in-migration at all adult ages. Underlying this pattern are very high rates of population turnover in line with the pattern observed generally across remote Australia (Taylor & Bell 1999). Furthermore, there has been stability in the shape of this non-Indigenous age pyramid over time, reflecting the ongoing role of this region within the Western Australian economy as a place of selective migration tied to short-term employment opportunity (Bell & Maher 1995). As a consequence, the Indigenous share of the regional population varies substantially according to age group, as Table 2.5 shows. Overall, the resident population of working age (15 years and over) amounts to 28 716, of whom 4256 (15%) are Indigenous. For the most part, among those under 20 years of age, Indigenous people account for between one-fifth and one-quarter of the regional population. However, this proportion drops in the working-age groups to between 10 and 20 per cent (or even lower for males) until the post-retirement years above 65 where the Indigenous proportion once again rises, especially among females.

Table 2.5. Indigenous percentage of five-year age groups: Pilbara SD, 2001

Age group	Males	Females
0–4	19.1	20.1
5–9	20.3	20.3
10–14	23.5	23.1
15–19	22.3	24.7
20–24	18.6	21.4
25–29	15.5	15.6
30–34	12.7	14.1
35–39	11.1	12.9
40–44	9.0	12.3
45–49	10.0	15.2
50–54	8.6	14.0
55–59	10.2	15.7
60–64	9.9	27.1
65 +	25.0	39.8
Total	19.1	23.4

The effect of this relative distribution by age and sex is reflected in the difference between Indigenous and non-Indigenous sex ratios as shown in Fig. 2.5. These are shown for both the ERP and place of enumeration (PoE) census counts. Thus, using both data sources, the Indigenous sex ratio is fairly close to parity but slightly in favour of males up to around age 35, after which point it tends to fall below parity, especially after age 55 due to relatively higher mortality among Indigenous males.

Figure 2.5. Indigenous and non-Indigenous sex ratios by age: Pilbara SD, ERP 2001

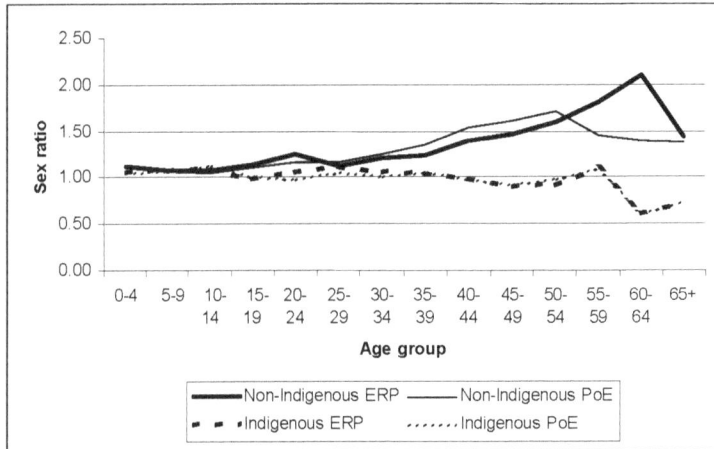

The non-Indigenous sex ratio is quite different with males predominant at all ages, especially over 30 years. Also of interest is the added difference evident between the non-Indigenous ERP and census counts, with males in the ERP profile more prominent in later ages over 50 years, and those in the census count profile more prominent in middle ages between 30 and 50 years, but declining as a ratio thereafter. This latter observation is likely to reflect the male-based composition of temporary workers in the 30–50 years age group.

Population projections

To date, policy development involving Indigenous populations has typically been reactive to needs as they become revealed (e.g. in terms of post-facto responses to housing shortages or employment needs), as opposed to being proactive in seeking to anticipate and plan for expected requirements. However, being proactive requires a measure of future requirements for infrastructure, programs, and services – a practice that is standard procedure for mainstream regional planning, and not least for mining industry business units. However, it is something that is rarely achieved, or even attempted, for Indigenous communities, where the approach to regional or settlement planning is much more prospective.

For example, State and local government planning authorities routinely develop future scenarios and often seek budgetary allocations on the basis of anticipated needs. A key element in this process is the production of small-area population projections or forecasts. While the ABS provides official projections of State and Territory and SLA populations, the individual States and Territories, in turn, also produce regional and local area projections, often down to the Local Government Area level (Bell 1992; Western Australian Planning Commission

2005). For these purposes a standard cohort-component methodology is generally applied, and this practice is adopted here to project the Indigenous population of the Pilbara to 2016. Ideally, population projections for a region such as the Pilbara that experiences major shocks to its regional economy would attempt to account for these and methods are available using input-output techniques and simple demographic-economic impact forecasting (Phibbs 1989). However, these require whole-of-region data input, especially in regard to workforce demand and composition (for example in terms of the FIFO component), and this task is a major enterprise by itself. As for the Indigenous population, however, relative detachment from the mainstream economy leaves it less susceptible to demographic-economic impacts and so a standard cohort-component methodology remains applicable.

Indigenous population projection assumptions

The cohort-component method carries forward the 2001 Indigenous ERP to 2016 by successive five-year periods. The projection is based simply on ageing the population by five-year blocs, subjecting each group to age- and sex-specific mortality, fertility and net migration regimes according to the following assumptions:

- Survival rates from the latest official Indigenous life tables for Western Australia (ABS 2004) are applied and held constant for the projection period. This assumption is consistent with evidence that life expectancy generally for Indigenous people has shown little sign of improvement in recent times (Kinfu & Taylor 2005).
- Age specific fertility rates (ASFRs) based on births to Indigenous women in the Western Australian Midwives Notification System for the Pilbara SD are applied. These data produce a Total Fertility Rate (TFR) of 2.9, which is substantially higher than the Indigenous TFR of 2.2 for Western Australia as a whole and is more in line with rates reported from similar remote regions of northern Australia (Kinfu & Taylor 2005). An additional fertility component is provided by Indigenous births to non-Indigenous mothers. This is estimated from registered births data supplied by the ABS to produce a partial total paternity rate of 0.31. As indicated in Fig. 2.6, Indigenous women's fertility remains relatively high at teenage years and peaks relatively early. Both of these ASFRs are held constant for the projection period, although the indications from fertility trend data for Western Australia as a whole would suggest that the former be allowed to decline over the time, and that the latter should rise. In the absence of any model as to how this might apply in the Pilbara region, both have been held constant.
- Migration is the most difficult to measure and yet most crucial component of regional population change in the sense that it has the potential to have the greatest demographic impact. One complicating issue for the Indigenous

population is the prevalence of short-term circular movement in the overall context of total mobility (Taylor 1998; Taylor & Bell 2004), although in many remote regions, such as the Pilbara, the same can be said for the non-Indigenous population as well (Bell 2001; Bell & Ward 2000).

• The indication from the ERP age distribution, with its loss of population in the 15–24 years age group, is that net migration does play an important role in shaping the composition of the Pilbara Indigenous population with consequences for future growth. In order to explore a means of capturing this effect, age- and sex-specific net migration rates were calculated using five-year inter-regional migration data from the 2001 Census. The result, shown in Fig. 2.7, yields a highly confused and counter-intuitive pattern of movement with net gains of children up to 19 years, and net losses for most adult years up to age 49, followed by substantial gains and losses among older people. Overall, the net effect is a migration gain of around 130 persons. However, in the previous intercensal period the Indigenous population of the Pilbara experienced net migration loss of a similar magnitude (Taylor & Bell 1999). In the absence of any logical pattern to these migration flows, and with no operational model of migration in and out of the region, these data are excluded from the calculation and migration rates are simply held constant at zero for all ages over the projection period.

• Finally, no allowance is made for population change via shifts in Indigenous identification.

Figure 2.6. Total fertility and partial paternity rates: Indigenous males and females in the Pilbara SD, 2001

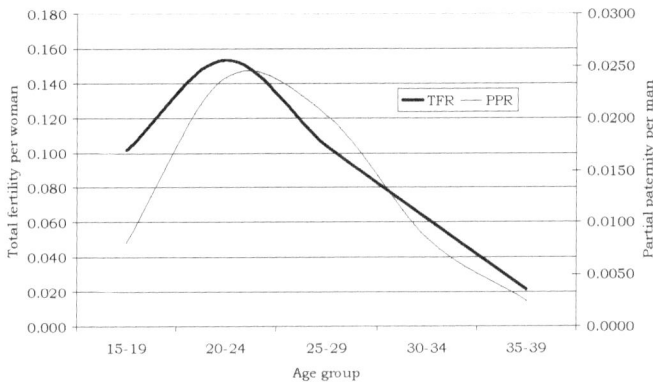

Source: Customised ABS tables, calculations by Yohannes Kinfu.

Figure 2.7. Age- and sex-specific pattern of Indigenous net migration: Pilbara SD, 1996–2001

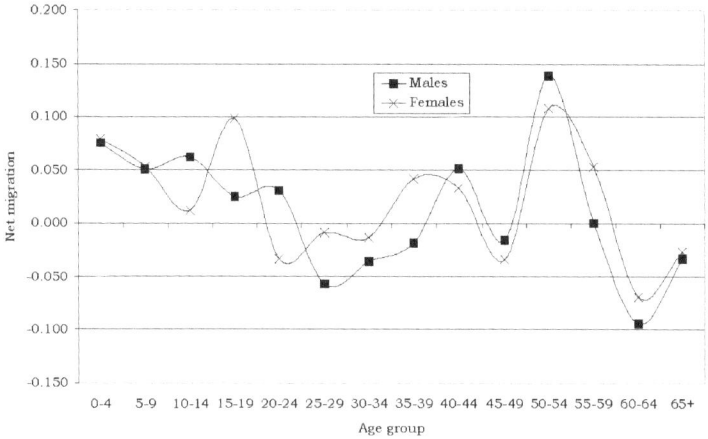

Source: Customised ABS tables, calculations by Yohannes Kinfu.

Against these parameters, the projection is conducted separately for males and females in five-year blocs from 2001 to 2016. Projected births for the 2001–06 period are added to the existing 2001 population and each cohort is then subjected to respective survival and net migration rates to arrive at an estimate of the population in each age group in 2006. This process is continued through to 2016.

As for projections of the non-Indigenous population, the simple approach adopted here is to derive these as a residual between the Indigenous projections and projections of the total regional population. Projections for the total population have been produced by the Western Australian Planning Commission using 2001 ERPs as the base (Western Australian Planning Commission 2005). However, given that the assumptions underlying the development of estimates for the Indigenous and total populations are inevitably quite different, the creation of a residual (non-Indigenous) population in this way is statistically problematic. Any estimation and projection of a 'non-Indigenous' population would ideally need to be guided by its own unique underlying assumptions, and the development of these is beyond the scope of the present exercise. Indeed, the social construction of such a population raises questions as to whether it is statistically possible at all. Nonetheless, an ability to consider the likely changing balance of Indigenous to non-Indigenous population shares in the region is critical to the establishment of meaningful quotas and targets in such areas as labour market planning. While the residual method of estimation is therefore retained, any conclusions drawn are presented as a guide only.

Table 2.6. Indigenous population of the Pilbara region by five-year age group: 2001 and 2016

Age group	ERP 2001	Projection 2016	Net change	% change
0–4	720	1059	339	47.0
5–9	754	993	239	31.7
10–14	784	906	122	15.6
15–19	600	712	112	18.6
20–24	537	738	201	37.5
25–29	601	756	155	25.8
30–34	555	569	14	2.6
35–39	474	499	25	5.3
40–44	368	542	174	47.4
45–49	344	484	140	40.6
50–54	240	393	153	63.9
55–59	162	285	123	76.2
60–64	118	241	123	104.7
65–69	257	336	79	30.9
Total	6514	8515	2001	30.7

Source: Based on calculations by Yohannes Kinfu and authors.

Indigenous population totals projected from 2001 to 2016 for the Pilbara region are shown in Table 2.6 by five-year age group, together with numeric and percentage change from the 2001 ERP. Overall, by 2016, the Indigenous population is projected to increase by 31 per cent to reach a population of 8515, an increase of some 2000 persons. As shown by the percentage change in each age group, much of this growth will occur in later years over the age of 50, while numerically there are large gains at younger ages. However, shifts in the size of individual five-year age groups are as much a function of the shape of the base ERP population and the overall ageing of prior cohorts. Some of the variability in growth rates by age group can be ironed out by aggregation. This also has the advantage of focusing on ranges that typically form the target of policy intervention, as shown in Table 2.7.

Table 2.7. Distribution of the Pilbara Indigenous population by select age groups: 2006–2016

Age group	2006	2016	Change (no)	Change (%)
0–4	911	1059	148	16.2
5–14	1470	1899	429	29.2
15–24	1371	1450	79	5.8
25–54	2769	3244	475	17.1
55+	619	863	244	39.4
Total	7141	8515	1372	19.2

Source: Based on calculations by Yohannes Kinfu and authors.

The selection of age groups is dictated somewhat by the availability of ERP data at five year intervals only. Thus, the infant years leading up to compulsory schooling are identifiable as 0–4 years, but for the years of compulsory schooling we are forced to use 5–14 years. Thereafter, we can identify the transition years from school to work as ages 15–24 years, while the prime working-age group is identified here as ages 25–54. Typically in the Australian workforce, and in International Labour Organisation convention, working age extends to 64 years with those over 65 years representing the aged and pensionable. However, given the evidence for premature ageing in the Indigenous population in the context of high levels of adult mortality and morbidity (Divarakan-Brown 1985; Earle & Earle 1999), the top of the working-age range has been set here at the earlier age of 55 years.

In Table 2.7, rather than report these data for the base year (2001), the projected year (2006) has been selected as the base in order to focus on likely growth in these cohorts from the present into the immediate years ahead. The results indicate an imminent infant population of some 900 accounting for almost 13 per cent of the regional total, while the 'school-age' population of almost 1500 is just over one-fifth of the regional population. Those in the transition years from school to work number almost 1400, again almost one-fifth of the population, while the working-age group of almost 2800 comprises more than one-third of the total and is the largest single social policy grouping. By comparison, the aged population is relatively small, even given the lower age at which group is set to start. If we look at how these groups are projected to grow over the next ten years, we observe the greatest numeric increase among those of prime working age, followed by those of school age. The other feature is the relatively large proportional increase in the aged population.

As for projections of the total Pilbara population, these have been being prepared by the Western Australian Planning Commission. They indicate a population of 42 900 by 2006 – a figure that is somewhat above the ABS ERP for the Pilbara in 2004 of 39 311. It is to be expected that the recent expansion of mining and related infrastructure development in the region will lead to enhanced population growth. Thus, a total of 21 development projects valued at $27.3 billion were committed or being considered for investment in the Pilbara at the end of 2003 (Argus Research 2004: 7). Not surprisingly, the biannual labour market forecasts produced by the Centre of Policy Studies, Monash University (December 2004 version) indicate a 12 per cent expansion in overall numbers employed in the Pilbara from 22 241 in 2003–04 to 24 957 by 2011–12. However, the Monash model is essentially a top-down perspective of employment change and while this provides the advantage of locating the Pilbara within a defensible view about the future of the Australian economy, it has only limited capacity to incorporate local bottom-up information which severely constrains its application in a region as dynamic as the Pilbara. Accordingly, any estimates from this source are used cautiously, and almost certainly refer to conservative minimum levels.

As for what this sort of expansion might imply for total population growth based on population multipliers, a crude way of establishing some measure of this is by assuming that the employment to population and working-age to total population ratios remain constant throughout the forecast period. So, with an estimated regional employment to population ratio of 78.7 in 2004, and a working-age share of population of 72.7, the Monash employment forecast figure for 2011–12 would imply a total population by then of around 43 600. This is not inconsistent with the Planning Commission projection of 44 400 by 2011, and given that the Monash employment forecast may well be conservative it lends some credence to the Commission's figures.

Thus, if we adopt the Planning Commission's projections for the total population on the strength of this corroborative evidence, we can now derive an estimate of the projected non-Indigenous population to continue on the historic series in Table 2.4 by assuming that this is the residual from projections of the Indigenous and total populations. As shown in Table 2.8, compared to the recent decline shown in Table 2.4, the non-Indigenous population is set to increase over the next decade to reach almost 40 000 by 2016. However, this still represents a lower rate of growth than projected for the Indigenous population, and so the Indigenous share of the regional population will continue to rise, reaching 18 per cent by 2016.

Table 2.8. Projected non-Indigenous population: Pilbara SD, 2006–2016

	2006	2011	2016
Total	42 900	44 400	46 600
Indigenous	7141	7817	8515
Non-Indigenous	35 759	36 583	38 085

Sources: Total projections from Western Australia Department for Planning and Infrastructure; Indigenous population projections based on calculations by Yohannes Kinfu and authors.

While these projections are correct according to the algorithms applied, they are only preliminary and there are several refinements that, if developed, would provide for greater certainty in the assumptions. In particular, there may be scope for some refinement of net migration assumptions if we had a greater appreciation of the general social and economic factors that may induce migration. An allied issue here would be more detailed analysis of inter-regional population movement for education and training purposes. Also, as noted, a more customised mortality profile is under construction and may be applied.

One device frequently deployed to canvass a range of possible projection outcomes is the calculation of several projection series based on varying assumptions. The current calculations involve the use of only one series. An obvious option, then, for further development of these projections would be to generate alternative scenarios based on possible combinations of falling/rising/stable fertility and mortality and varying assumptions about net migration. While there is some heuristic potential here, it obviously makes sense to base such exploration on plausible indicators, and so the indicators themselves would also need to be assessed as part of a longer-term project.

Finally, in using the projections as a means of targeting policy, it is possible to estimate the future size of the Indigenous and non-Indigenous resident labour force by applying labour force participation rates to the projected working-age populations. If select employment/population ratios are also applied, then the quantum of need for additional job creation can also be calculated according to specified or agreed employment levels. This exercise essentially represents a regionalised version of similar calculations of Indigenous employment demand made at the national level (Hunter & Taylor 2004; Taylor & Hunter 1998), and it is to this task that we now turn.

3. Indigenous participation in the regional labour market

Indigenous participation in the Pilbara labour market is long-standing and stems from the first incursion of pastoralists and miners into the region in the latter part of the nineteenth century (Holcombe 2004; Wilson 1980). Within more recent history, however, the significant development was the eviction of Indigenous labour from the pastoral industry, a process that commenced with the declaration of the equal pay award in 1968 (Edmunds 1989: 28). While this shedding of labour has yet to be replaced by any firm engagement of Indigenous people in the regional labour market, the current expansion of mining activity combined with the framework of agreement-making and corporate resolve to employ Indigenous workers has the potential (in terms of scale at least) to offer unprecedented opportunity for enhanced Indigenous participation. While this current impetus is generated by mining and related activities, it should be remembered that the contemporary expansion of the regional labour market has also been associated with the growth of tourism, provision of government services, and the Indigenous community sector.

The rich and diverse history of Indigenous engagement with the labour market reveals a population achieving highly varied outcomes, balancing pressures to survive in the modern economy with the need and desire to retain culture, and moving, as a consequence, between many employment niches in all sectors including mining, pastoralism, private business and community organisations (Olive 1997). Others saw themselves as simply excluded from opportunity (see Interview segment 6, p. 58; Interview segment 13, p. 61), a view supported by more formal analyses of the historic record (Cousins & Nieuwenhuysen 1984: 131–40). Whatever the circumstances, the net effect, despite substantial growth in economic activity and employment opportunity in the Pilbara since the 1960s, is that the overall employment rate for Indigenous people rose only slightly from 38 per cent of all adults in 1971 to just 42 per cent in 2001. This compares with equivalent figures for non-Indigenous adults in the Pilbara of 81 per cent and 80 per cent respectively. Thus, over the past 30 years, the rate of Indigenous employment in the Pilbara has risen from just less than half of the non-Indigenous rate to just slightly over half. In terms of numbers, non-Indigenous employment has risen from 16 352 in 1971 to 19 671 in 2001, while Indigenous employment increased from 814 to an estimated 1808. However, much of the growth in Indigenous employment, certainly through the 1990s, was due to increased participation in the Community Development Employment Projects (CDEP) scheme. It is therefore necessary to adjust for this in any discussion of employment gains. Before reviewing the present composition of the regional workforce it is also necessary to establish some basic population parameters.

Given the nature of the regional economy and its part-dependence on a peripatetic workforce, it is necessary to be clear about which workforce population is being referred to in any analysis. This is because of the distinction between individuals who consider the Pilbara to be their usual place of residence, as opposed to those who might be counted in the Pilbara on census night but whose usual place of residence is actually elsewhere. This latter group can include many FIFO workers as well as others who acquire short-term contract work, or who service the region from an outside base. While such individuals experience high turnover, as a group they comprise a vital and constant structural component of the regional labour market (Storey 2001). Any discussion of workforce levels and composition must, where appropriate, include such elements.

While this much seems unequivocal, accurate data on this mobile workforce for the region as whole is difficult to compile as this would ideally require the bringing together of disparate company records. As a fall-back, census data can be used to estimate the size of the temporary workforce. Thus, in 2001, the number of employed adults counted in the Pilbara on census night was 13 per cent greater than the number of workers who indicated that the Pilbara was their usual place of residence. At face value, then, it might be suggested that any difference between population counts by employment status provides for an indirect measure of the temporary workforce. On this basis, the size of the census-derived temporary workforce in 2001 would have been around 2500, or 11 per cent of the total. Virtually all of these would have been non-Indigenous workers since Indigenous usual residence counts and place of enumeration counts were very similar.

While the regional labour market has grown in both size and complexity, Indigenous participation has remained relatively marginal. In effect, the past 30 years have witnessed a shift from an almost total reliance on the private sector for employment (mostly in the pastoral industry), to increased reliance on the government sector in the form of CDEP and the community services industry. Beyond this, as noted, only 30 per cent of Indigenous adults participate in the mainstream labour market compared to 80 per cent of non-Indigenous residents, and a rising share of this mainstream employment is now in the mining industry. This structural gap in employment, together with overall low levels of Indigenous labour force participation, has significant consequences for current Indigenous economic status.

In this context, one question that looms large is whether the targets set by Pilbara Iron and other companies for recruiting Indigenous labour have the capacity to lead to improvement in overall regional labour force and economic status. This is not to suggest that such goals should necessarily be met by mining company employment strategies alone, rather that even despite these strategies it is possible that effecting overall change in regional labour force status may prove to be intractable. The essential background to considering this question is one of projected growth in the Indigenous working-age population set against likely future employment demand and labour supply. To explore this we begin by estimating current levels of Indigenous labour force status.

Regional labour force status: rates and estimated levels

Rates of labour force status are presented in Table 3.1. for Indigenous and non-Indigenous residents of the Pilbara, using 2001 Census usual residence counts. Also provided are estimates of 2001 labour force status levels derived by applying the census-derived rates to the respective Indigenous and non-Indigenous adult ERPs. This compensates for census undercount. Three standard indicators of labour force status are established:

- the employment to population ratio, representing the percentage of persons aged 15 years and over who indicated in the census that they were in employment (either in CDEP or mainstream work) during the week prior to enumeration;
- the unemployment rate, expressing those who indicated that they were not in employment but had actively looked for work during the four weeks prior to enumeration, as a percentage of those aged 15 years and over;
- the labour force participation rate, representing persons in the labour force (employed and unemployed) as a percentage of those of working age.

Table 3.1. Labour force status rates and estimated levels[a] for Indigenous and non-Indigenous residents of the Pilbara, 2001

	Employed		Unemployed	Not in the labour force	Total 15 +
	CDEP	Mainstream			
Census rates					
Indigenous	12.3	30.2	8.0	49.5	100.0
Non-Indigenous	0.1	80.3	3.0	16.6	100.0
Estimated levels					
Indigenous	524	1284	339	2108	4256
Non-Indigenous	20	19 651	731	4058	24 460

[a]Based on ERP.

Overall, in 2001, an estimated 1808 Indigenous adults were employed, with 71 per cent of these in mainstream work. From these data, a further 339 were unemployed and actively seeking work, while 2108 were not in the labour force. Interestingly, this estimate of unemployed numbers is far lower than the 670 Indigenous Centrelink clients in the Pilbara who were in receipt of Newstart payments in 2005. Even though these data refer to different points in time, this discrepancy is significant after accounting for variation in the definition of unemployment in the two collections. Why such a gap should occur is not known. One possibility is that many Newstart clients were exempt from activity testing and may have been recorded under 'CDEP' or 'Not in the Labour Force' in census data. At the same time, the estimate of 524 employed in CDEP is way below the figure of 1032 Indigenous participants recorded by ATSIC in Pilbara CDEP schemes at the time of the 2001 Census. Once again, it is not known why this discrepancy occurs, but it may reflect the fact that the census records employment in the 'last' week, while CDEP work is intermittent and predominantly part-time.

Indeed, given the administratively-determined nature of much Indigenous economic activity in the region, the boundaries between officially recorded employment, unemployment, and consequent labour force participation, are sufficiently blurred to require approaching all these data with some caution. They are best seen as rough estimates rather than as robust measures. Of interest, though, is the fact that each of the census-based Indigenous labour force status rates for the Pilbara are almost identical to the Western Australia Indigenous state average. It should also be noted that current (2005) figures for the five CDEP schemes in the region based on administrative data indicate a total of 550 participants, which is very close to the census-based estimate derived above.

It is clear that the Indigenous labour force is much smaller than the non-Indigenous labour force, and disproportionately so as Indigenous adults comprised only 9.5 per cent of the resident regional labour force despite accounting for 15 per cent of the resident adult population. Of course, if we now add the estimated 2500 non-Indigenous non-residents who are employed in the region on temporary arrangements as in our previous calculation, then Indigenous people would account for an even lower share of the effective regional labour force.

Of particular interest for social impact planning is the distribution of labour force status by age. This is shown in Fig. 3.1 for Indigenous adults in the Pilbara with the actual rates provided in Table 3.2. The most striking feature is that those outside of the labour force comprise the largest single category in all but the 35–44 age group. This is especially noticeable among young adults aged 15–24, half of whom are outside of the labour force. The main reason for this at older ages is a steady falling off in labour force participation beyond age 54, with mainstream employment declining substantially. Thus, those most active in the labour market are in the fairly narrow age range of 35–44 years. As far as CDEP is concerned, there does appear to be some relationship between participation and age with rates highest among those aged 15–24 and declining steadily thereafter.

Figure 3.1. Labour force status rates by age group: Indigenous adults in the Pilbara, 2001[1]

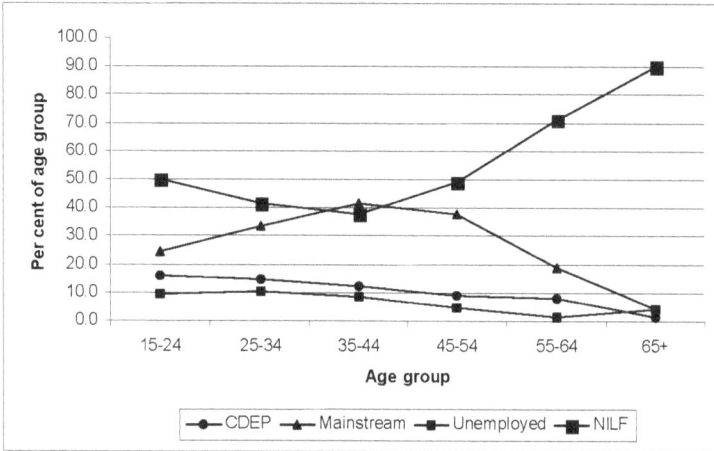

Source: ABS 2002b.

Table 3.2. Labour force status rates by age group: Indigenous adults in the Pilbara,[a] 2001

	CDEP	Mainstream	Unemployed	NILF[b]	Total
15–24	15.9	24.7	9.6	49.8	100.0
25–34	14.7	33.7	10.3	41.3	100.0
35–44	12.4	41.5	8.6	37.6	100.0
45–54	8.8	37.6	4.7	49.0	100.0
55–64	8.1	19.0	1.4	71.4	100.0
65 +	1.4	4.3	4.3	89.9	100.0

[a]Based on place of enumeration

[b]NILF = not in the labour force.

Source: ABS 2002b.

Dependency ratios

Measures of the potential economic implications of a given age structure can be combined with data on labour force status to produce a range of dependency ratios. These are shown in Table 3.3 for the Indigenous population of the Pilbara in 2001, with comparisons drawn from Western Australia as a whole. The *childhood dependency* ratio is the simplest of these measures and expresses the number of children in the population (aged 0–14 years) as a ratio of the working-age population (defined here as aged 15–55 given the significance of premature adult mortality). A ratio of 1.0 would indicate that the size of the two age groups is the same and that there is one person of working age for every child. A figure greater than 1.0 indicates more than one child to each person of working age, and less than 1.0 indicates less than one child to each person of

[1]Based on place of enumeration.

working age. Obviously, this only provides an indication of potential economic providers to dependents as it takes no account of the economically inactive.

Table 3.3. Dependency ratios for the Indigenous population of the Pilbara and Western Australia,[a] 2001

Dependency ratio	Pilbara	Western Australia
Childhood dependency	0.6	0.7
Childhood burden	1.2	1.6
Childhood burden (excl. CDEP)	1.8	2.3
Dependency ratio	2.0	2.2
Economic burden	2.6	2.9
Economic burden (excl. CDEP)	3.7	4.8

[a]Based on place of enumeration.
Source: ABS 2002b.

In the Pilbara, the childhood dependency ratio was 0.6 which is broadly similar to the 0.7 reported for Indigenous people generally in Western Australia. In effect, there are 0.6 Indigenous children to each Indigenous adult of working age. While this may appear to be a favourable ratio at one level, it represents far more children per adult compared to the ratio of 0.3 recorded for the non-Indigenous population of the region.

More refined measures of dependency attempt to incorporate some indication of the ability of working-age adults to support others. The childhood burden, for example, is defined as the ratio of the number of children to the number of employed persons. Once again, a figure of 1.0 indicates parity. According to census-based estimates, there were 1.2 Indigenous children to each employed adult if all those engaged by the CDEP scheme are considered to be in employment. However, if this calculation is based on those employed only in non-CDEP work the ratio rises to 1.8. The fact that the equivalent ratio for all Indigenous people in Western Australian was higher again at 2.3 underlines the fact that non-CDEP work is a relatively important support mechanism for large numbers of Indigenous child dependents in the Pilbara.

Another measure is provided by the dependency ratio, which represents the ratio of children and economically inactive adults to adults in the labour force (those employed plus those unemployed). This produces an average of 2.0 dependents per economically active person, although if the focus were solely on those in mainstream employment the dependency ratio would be much higher at 3.4.

Finally, the economic burden is a ratio of the number of children and economically inactive persons (including here those unemployed) to employed persons. This shows that for each employed Indigenous person (including those in the CDEP scheme) there are 2.6 other Indigenous people who are not employed, a figure close to the state average. If, however, those in CDEP are excluded from the economically active then the economic burden in the Pilbara rises to 3.7 dependents per income earner, although this is still below the state average.

From a regional planning perspective, then, the youthful Indigenous age profile is a key demographic feature when set against the relatively poor labour force status of Indigenous adults. In effect, there are almost four dependents, on average, for each Indigenous employee in the mainstream labour market. In the local context of access to resources and consumer spending, this represents a notably higher economic burden for the regional Indigenous population than recorded for non-Indigenous residents (0.6 dependents per income earner) with whom Indigenous residents can draw direct comparison.

Table 3.4. Indigenous Organisations in Pilbara SLAs, 2004

Port Hedland SLA

Pilbara Aboriginal Chamber of Commerce

Pilbara Meta Maya Regional Aboriginal Corporation

Pilbara Native Title Service/Yamatji Marlpa Maaja Aboriginal Corporation

Wangka Maya Pilbara Aboriginal Language Centre

Western Desert Purntupukurna Aboriginal Corporation

Jinparinya Aboriginal Corporation

Marta Marta Aboriginal Corporation

Mugarinya Community Association

Tjalka Boorda Community

Tjalka Waru Community

Bloodwood Tree Association (Inc)

IBN Corporation P/L (Foundation)

Marapikurrinya Aboriginal Corporation

Mulba Radio (Port Hedland Indigenous Media Aboriginal Corporation)

Ngarda Ngarli Yarndu Foundation

Pilbara Arts, Crafts and Designs Aboriginal Corporation

Pilbara Indigenous Women's Aboriginal Corporation

Port Hedland Regional Aboriginal Corporation

Port Hedland Sobering Up Centre Inc.

Punju Ngarugudi Njamal

Wirraka Maya Aboriginal Corporation-Aboriginal Health and Medical Service

Yandeyarra Aboriginal Pastoral Company Pty Ltd

Yandeyarra Aboriginal Store Pty Ltd

Hedland CDEP Aboriginal Corporation

Mugarinya Community Association (CDEP)

Pilbara Meta Maya Regional Aboriginal Corporation (CDEP)

Roebourne SLA

Mawarnkarra Health Service

Bumajina Aboriginal Corporation

Bujee-Nhoor-Pu Aboriginal Cultural Enterprise

Ieramugadu/Mt. Welcome Pastoral Company

Jukuwarlu Aboriginal Corporation

Manga Thandu Maya-Roebourne Safe House

Marra Gootharra Aboriginal Corporation

Mingga patrol

Minimurghali Mia Education Centre

Ngarluma/Yindjibarndi Foundation

Oondoomarra Ngarluma Aboriginal Corporation

Pilbara Aboriginal Church

Roebourne Sobering Up Shelter

Roebourne Youth Centre

Warba Mirdawaji Arts Group

Warawarni-Gu (Healing) Art

Yaandina Family Centre

Yathalla Group Aboriginal Corporation

Cheeditha Group Aboriginal Corporation

Mingulltharndo Aboriginal Corporation

Weymul Aboriginal Corporation

Ngarliyarndu Bindirri Aboriginal Corporation (CDEP Roebourne)

Ashburton SLA

Buurabalyi Thalanyji Aboriginal Corporation

Gumala Aboriginal Corporation

Jundaru Aboriginal Corporation

Walyun Mia – Onslow Safe House

Wanu Wanu

Bindi Bindi Community

Ngurawaana Aboriginal Corporation

Wakathuni Aboriginal Corporation

Bellary Springs

Youngalina Banyjima Aboriginal Corporation

Ashburton Aboriginal Corporation (CDEP Tom Price)

Ashburton Aboriginal Corporation (CDEP Onslow)

East Pilbara SLA

Irrungadji Group Association

Jigalong Community Inc.

Kiwirrkurra Community

Kunnawarritji Aboriginal Corporation

Mirtunkarra Aboriginal Corporation

Nomads Charitable & Education Foundation

Parrnngurr (Cotton Creek)

Pipunya Community Inc.

Parnpajinya Aboriginal Association

Punmu Aboriginal Corporation

Warralong Aboriginal Corporation

Industry sector

Private sector economic activity dominates the Pilbara labour market and accounts for as much as 81 per cent of locally-resident employees. As noted earlier, Indigenous employment was also mostly in the private sector up to the 1970s, but with structural change in the pastoral industry and limited intervening opportunities found in the mining sector, the trend for the Indigenous workforce has subsequently been towards greater reliance on the public sector for employment. As Table 3.4 implies, only half (49%) of Indigenous employees in the Pilbara are now engaged in the private sector. Almost one-third (31%) are now employed by CDEP, while state government employment accounts for 13 per cent the Indigenous workforce. This relatively distinct composition of employment by industry sector is reflected in the distribution of part-time and full-time work. Almost half of all Indigenous employment is part-time only (compared to just 27% overall), and a major reason for this is the reliance for much employment on CDEP.

Table 3.5. Indigenous employment by industry sector and hours worked:[a]
Pilbara SD,[b] 2001

	Number			Per cent		
	Part Time	Full Time	Total	Part Time	Full Time	Total
Commonwealth government	10	24	34	29.4	70.6	100.0
State government	79	103	182	43.4	56.6	100.0
Local government	18	28	46	39.1	60.9	100.0
Private sector	222	456	678	32.7	67.3	100.0
CDEP	350	86	436	80.3	19.7	100.0
All workers	679	697	1376	49.3	50.7	100.0

[a]Excludes industry sector and hours worked not stated

[b]Based on place of enumeration.

Source: ABS 2002b.

Aside from CDEP, this portrayal of Indigenous employment by industry sector masks a very important component of the Indigenous labour market that has been labelled elsewhere as the Indigenous community organisation sector (Rowse 2002). This sector is significant not only for its growth over the past three decades, but also for the fact that employment levels in Indigenous community organisations have invariably been counter to economic cycles as they are dependent more on government funding regimes and the flow of localised private sector monies, not least based around such initiatives as mining agreements. In the 1994 National Aboriginal and Torres Strait Islander Survey (NATSIS), an estimated 11 per cent of Indigenous people who were employed in the Ngarda Ngarli Yarndu ATSIC Region reported that they were employed by an Indigenous community organisation. The equivalent proportion in the Warburton ATSIC Region (the northern half of which falls within the Pilbara SD) was 58 per cent. While no more recent similar data exist, Table 3.5 provides a list of Indigenous organisations in the Pilbara in 2004 in order to provide a measure of the likely scale and scope of employment in this sector. If each of these employed just three Indigenous workers, this would amount to more than the Commonwealth and Western Australian Government sectors combined.

Source: Pilbara Development Commission.

Industry and occupation

In the final analysis, employment is a means to personal income generation, with the amount generated is determined largely by occupational status. In turn, the availability of particular occupations within the region is partly related to the industry mix of economic activities. Thus, the relative distribution of Indigenous and non-Indigenous employment by industry and occupational category is a vital feature of participation in the regional labour market and this is shown in Figs. 3.2 and 3.3 for male and female workers respectively.

Figure 3.2. Distribution of Indigenous and non-Indigenous male employment by industry division: Pilbara SD, 2001

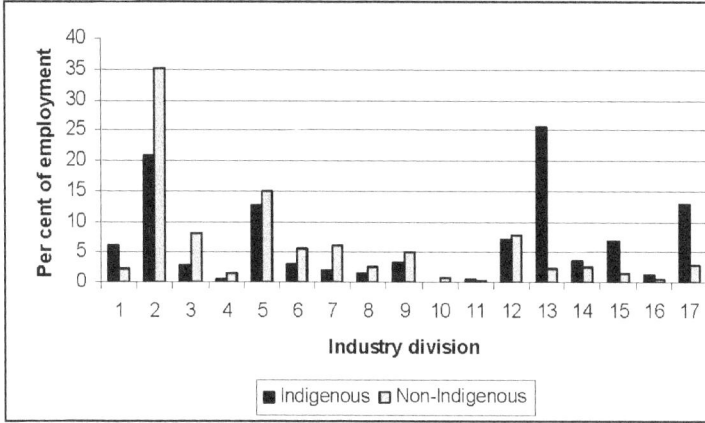

Source: ABS 2001 Census customised tables.

Key: 1.Agriculture, Forestry and Fishing; 2. Mining; 3. Manufacturing; 4. Electricity, gas and water; 5. Construction; 6. Wholesale Trade; 7. Retail Trade; 8. Accommodation, cafes and restaurants; 9. Transport and Storage; 10. Communication Services; 11. Finance and Insurance; 12. Property and Business Services; 13. Government Administration and Defence; 14. Education; 15. Health and Community Services; 16. Cultural and Recreational Services; 17. Personal and Other Services.

Figure 3.3. Distribution of Indigenous and non-Indigenous female employment by industry division: Pilbara SD, 2001

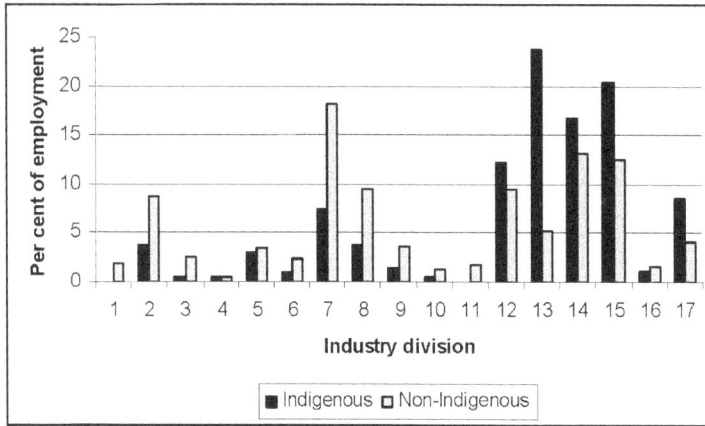

Source: ABS 2001 Census customised tables.

Key: 1.Agriculture, Forestry and Fishing; 2. Mining; 3. Manufacturing; 4. Electricity, gas and water; 5. Construction; 6. Wholesale Trade; 7. Retail Trade; 8. Accommodation, cafes and restaurants; 9. Transport and Storage; 10. Communication Services; 11. Finance and Insurance; 12. Property and Business Services; 13. Government Administration and Defence; 14. Education; 15. Health and Community Services; 16. Cultural and Recreational Services; 17. Personal and Other Services.

The overwhelming importance of employment in the mining industry for non-Indigenous males is clearly demonstrated with direct employment in the industry accounting for more than one-third of the regional non-Indigenous male workforce. Mining is also of significance for Indigenous males, though it is outweighed as the primary employer by government administration (which mostly reflects ABS census coding for CDEP work). Construction industry employment (much of it tied to mining activity) is also prominent for both Indigenous and non-Indigenous males, while something of a historical legacy is evident in the greater representation of Indigenous males in agriculture and their relative absence from manufacturing. These variations contribute to relatively high indices of dissimilarity, especially for males. In short, if the Indigenous workforce were to participate in the industry mix of the regional labour market in the same fashion as non-Indigenous workers, then according to the index of dissimilarity more than one-third of them (38%) would need to change their industry of employment. Obviously, this would represent a substantial restructuring. The pattern of regional employment appears to be much more segregated between Indigenous and non-Indigenous women. The former are heavily concentrated in government administration (CDEP), education, and health and community services, and are notably absent, when compared to their non-Indigenous counterparts, from mining, retail trade, accommodation, and transport and storage industries.

A similar scale of difference in workforce participation is evident in respect of occupational distributions (Figs 3.4 and 3.5). Among males, the overwhelming pattern is one of under-representation of Indigenous workers in managerial, professional and trade occupations, and their substantial over-representation in labouring jobs. While much of the latter arises from the ABS tendency to code CDEP scheme workers as labourers and related workers, the contrasting distributions focused on either end of the occupational scale provide one measure of the skills differential between Indigenous and non-Indigenous males. The pattern of female occupational distribution is less diverse, although it is true that Indigenous women experience double segregation in the labour market since (as women) they gravitate to the same sex-segregated occupations as their non-Indigenous counterparts. However, Fig. 3.5 shows that Indigenous women are more heavily concentrated than other females in intermediate level clerical jobs while the relative proportions of para-professionals are reasonably close, certainly more so than among men. This is reflected in a lower index of dissimilarity between female workers compared to male (24.2 compared to 34.0).

Figure 3.4. Distribution of Indigenous and non-Indigenous male employment by occupational group: Pilbara SD, 2001

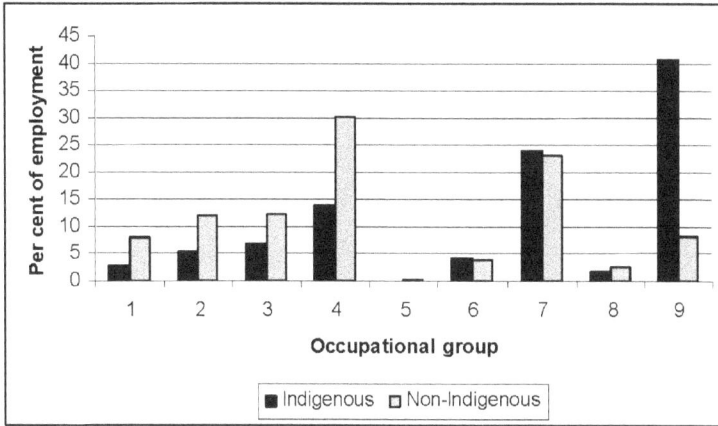

Source: ABS 2001 Census customised tables.

Key: 1.Managers and Administrators; 2. Professionals; 3. Associate Professionals; 4. Tradespersons and Related Workers; 5. Advanced Clerical and Service Workers; 6. Intermediate Clerical, Sales and Service Workers; 7. Intermediate Production and Transport Workers; 8. Elementary Clerical, Sales and Service Workers; 9. Labourers and Related Workers

Figure 3.5. Distribution of Indigenous and non-Indigenous female employment by occupational group: Pilbara SD, 2001

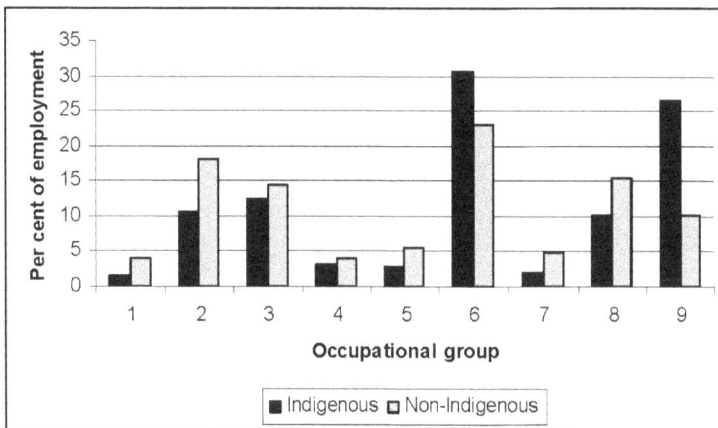

Source: ABS 2001 Census customised tables.

Key: 1. Managers and Administrators; 2. Professionals; 3. Associate Professionals; 4. Tradespersons and Related Workers; 5. Advanced Clerical and Service Workers; 6. Intermediate Clerical, Sales and Service Workers; 7. Intermediate Production and Transport Workers; 8. Elementary Clerical, Sales and Service Workers; 9. Labourers and Related Workers.

The data in Figs. 3.4 and 3.5 reveal only the broad outlines of the regional labour market. Each of these classifications can be disaggregated into more detailed descriptions of industry and occupation in a way that highlights the particular jobs that Indigenous and non-Indigenous workers congregate in. For example, the Australian and New Zealand Standard Industrial Classification (ANZSIC) that the ABS uses to categorise industries can be broken down into 635 individual industry classes, while the Australian Standard Classification of Occupations (ASCO) is reduceable to 340 occupational unit groups. In identifying key components of the Pilbara labour market, both ANZSIC and ASCO categories utilised here.

When examined at this level of detail, the distribution of employment in the Pilbara, for both Indigenous and non-Indigenous workers emerges as even more concentrated into relatively few individual industries and occupations. Table 3.6 shows the top 20 industry classes (based on numbers employed) listed in rank order for both sets of workers. These top 20 out of 635 industries account for as much as two-thirds of all Indigenous employment, and approaching half (49%) of all non-Indigenous employment. Industries marked in bold indicate those that are common to the Indigenous and non-Indigenous lists; all others are unique to one list or the other. Thus, half of the top 20 employing industries are common to both Indigenous and non-Indigenous workers and, not surprisingly, many of these (such as iron ore mining, supermarkets, primary education, and hospitals) are major regional employers.

However, there are notable differences indicating significant structural breaks in the labour market. For example, Indigenous people are absent from certain components of mining and trade-based industries, as well as from key elements of the tourism sector such as cafes and restaurants. By contrast, they are more likely to be found in municipal-type service industries such as non-residential care, gardening, and waste disposal services, some of which are the result of entry-level employment and training programs sponsored by mining companies. Examples include the Ieramugadu Corporation at Roebourne that has provided contract gardening maintenance and clean-up services to Pilbara Iron port operations at Dampier, and the Martu Contract Gardening project at Newman developed by BHP Iron Ore and the Western Desert Puntukurnuparna Aboriginal Corporation. At a stroke then, the relative absence of Indigenous people from some of the region's top employing industries indicates that a significant contribution to the relatively poor labour force status of Indigenous people is their failure to achieve parity participation across the full range of activities associated with the region's key economic sectors.

Table 3.6. Rank order of top 20 industries of employment: Indigenous and non-Indigenous workers in the Pilbara SD, 2001

Indigenous	No. employed	Non-Indigenous	No. employed
Local Government Administration [a]	262	Iron Ore Mining	3515
Iron Ore Mining	113	Primary Education	551
Interest Groups, nec	59	Supermarket and Grocery Stores	526
Non-Residential Care Services, nec[b]	56	Accommodation	442
Primary Education	54	Non-Building Construction, nec	410
Cleaning Services	40	Cafes and Restaurants	351
Gardening Services	33	Hospitals	304
Central Government Administration	28	Electrical Services	272
Road and Bridge Construction	27	Secondary Education	271
State Government Administration	27	Metal and Mineral Wholesaling	265
Employment Placement Services	25	Oil and Gas Extraction	264
Waste Disposal Services	23	Road Freight Transport	257
Beef Cattle Farming	21	Site Preparation Services	241
Hospitals	21	Local Government Administration	237
Health and Community Services, undefined	20	Consulting Engineering Services	233
Site Preparation Services	19	Cleaning Services	230
Supermarket and Grocery Stores	18	Takeaway Food Retailing	225
Accommodation	18	Beef Cattle Farming	199
Health Services, undefined	18	Mining, nec	199
Community Services, undefined	17	State Government Administration	199
Total top 20 employment	1394	Total top 20 employment	18 887
% of workforce	64.5	% of workforce	48.7

[a]Shared categories in bold

[b]nec = not elsewhere classified.

Source: 2001 ABS Census of Population and Housing customised place of enumeration tables.

Segmentation and concentration in the regional labour market is also evident in regard to occupation. Table 3.7 reveals that the top 20 out of 340 occupations account for fully 55 and 40 per cent of Indigenous and non-Indigenous workers respectively, while the lists of occupations reveal significant differences. Only eight out of the top 20 occupational categories are shared in common, indicating greater occupational than industry segregation. While certain major occupations in the region are common to both populations, Table 3.7 also reveals a skill divide in occupational distribution. Thus, non-Indigenous workers are registered nurses and primary school teachers, whereas Indigenous workers are health workers and education aides. Labouring occupations do not appear in the non-Indigenous top 20, and professional/managerial occupations do not appear in the Indigenous list.

Table 3.7. Rank order of top 20 occupations of employment: Indigenous and non-Indigenous workers in the Pilbara SD, 2001

Indigenous	No. employed	Non-Indigenous	No. employed
Cleaners [a]	170	**Metal Fitters and Machinists**	1013
Education Aides	80	**Sales Assistants**	706
Garbage Collectors	52	**Miners**	689
Nursery and Garden Labourers	51	Electricians	591
Farm Hands	41	**Cleaners**	564
Mobile Construction Plant Operators	38	**Truck Drivers**	399
Metal Fitters and Machinists	37	Structural Steel and Welding Tradespersons	373
Labourers and Related Workers, nfd[b]	37	Other Building and Engineering Associate Professionals	348
Gardeners	34	Primary School Teachers	301
Truck Drivers	32	Shop Managers	290
Miners	32	**General Clerks**	287
General Clerks	26	Secretaries and Personal Assistants	250
Welfare and Community Workers	25	Office Managers	246
Intermediate Machine Operators, nfd	24	Storepersons	246
Other Labourers and Related Workers, nfd	23	**Mobile Construction Plant Operators**	242
Construction and Plumber's Assistants	21	Production Managers	228
Aboriginal and Torres Strait Islander Health Workers	20	Motor Mechanics	228
Sales Assistants	20	**Project and Program Administrators**	227
Kitchenhands	20	Checkout Operators and Cashiers	225
Project and Program Administrators	18	Registered Nurses	217
Total top 20 employment	801	Total top 20 employment	7670
% of workforce	55.0	% of workforce	40.3

[a]Shared categories in bold

[b]nfd = not further defined

Source: 2001 ABS Census of Population and Housing customised place of enumeration tables.

CDEP activities

One drawback in relation to census-derived industry and occupational data is their tendency to apply blanket classification to CDEP scheme employment. As shown above, this results in a high concentration of Indigenous employment in government administration (especially local government), and as labourers. It is also the case that because of the employment substitution effect of CDEP, much work that is classified as CDEP actually covers a wider range of industry and employment categories than is apparent from census coding. Examples here might be CDEP work for a mining sub-contractor, or as a health worker, or teachers aide. The argument here is that census coding of CDEP can mask diversity in the pattern of Indigenous participation in the regional economy. Given the key role played by CDEP in terms of bolstering regional labour force status there is a need to establish and account for this diversity of economic activity and explore ways in which vital elements might more fully articulate with mining-based developments either via direct contracting, sub-contracting and/or joint venturing.

The temporary workforce

As noted earlier, an estimate of the size of the temporary workforce in the Pilbara can be derived by comparing employed persons counted in the Pilbara on census night against their usual place of residence. On this basis, temporary workers account for around 11 per cent of the total workforce, although their presence in the region appears to loom larger than this in the minds of some local Indigenous residents (see Interview segment 2, p. 57; Interview segment 7, p. 58; Interview segment 14, p. 61; Interview segment 21, p. 74). The main distribution of these workers according to their detailed industry and occupational classifications is shown in rank order in Table 3.8. As might be expected, the largest number of temporary workers is found in iron ore mining with FIFO arrangements no doubt accounting for many of these. This is also the case for other areas of the mining industry as well as with allied activities such as site preparation and consulting engineering services. The construction industry is also prominent in employing temporary workers, while areas associated with tourism and the pastoral industry also emerge. Not surprisingly, this industry mix is reflected in the sorts of occupations undertaken by temporary workers. Thus, miners, drillers, metal fitters and geologists are listed, as are electricians, carpenters and civil engineers. Interestingly, also in the mix are more service-oriented workers such as registered nurses, secretaries and sales assistants.

Table 3.8. Temporary workers by rank order of top 20 industry and occupational classifications: Pilbara SD, 2001

Industry class	Occupation unit
Iron Ore Mining	Metal Fitters and Machinists
Non-Building Construction	Electricians
Consulting Engineering Services	Mobile Construction Plant Operators
Beef Cattle Farming	Geologists and Geophysicists
Other Mining Services	Structural Steel and Welding Tradespersons
Site Preparation Services	Livestock Farmers
Electrical Services	Drillers
Accommodation	Building, Architecture, Surveying Assistant Professionals
Metal Ore Mining	Truck Drivers
Gold Ore Mining	Building and Engineering Professionals
Construction Services	Farm Hands
Road and Bridge Construction	Registered Nurses
Cafes and Restaurants	Miners
Copper Ore Mining	Carpentry and Joinery Tradespersons
Grain-Sheep and Grain-Beef Farming	Secretaries and Personal Assistants
Mineral Exploration	Structural Steel Construction Workers
Hospitals	Mining Support Workers/Driller's Assistants
Business Management Services	Mixed Crop and Livestock Farmers
Road Freight Transport	Sales Assistants
Contract Staff Services	Civil Engineers

Source: 2001 ABS Census of Population and Housing customised place of enumeration tables.

Mining employment

The nature of mining development in the Pilbara since the 1960s (essentially a series of inland mines linked by rail to coastal processing and shipping operations) has produced two broad regional networks based around the operations of Hamersley Iron and Robe (now Pilbara Iron), and BHP Iron Ore (now BHP Billiton). The first of these networks links mines at Brockman, Marandoo, Paraburdoo, Channar, Eastern Range, Mt. Tom Price, Yandicoogina, and West Angelas to Dampier and Cape Lambert, and the second network connects the Yarrie mine, the Yandi and Area C mines, along with Mt Whaleback and Jimblebar in the Newman area to Port Hedland.

Of course, other significant mining operations exist throughout the Pilbara region, including the Woodside, Chevron Texaco, and BHP installations offshore, but Pilbara Iron and BHP Billiton have long been the dominant employers in the region and remain so. According to the Western Australia Department of Industry and Resources (2004: 46–9) Pilbara Iron and BHP Billiton accounted for as much as 10 315 (69%) of the average of 14 900 employed in Pilbara-based mine sites and associated operations in 2003–04. Interestingly, according to these estimates, these two companies had almost equivalent average workforces (5187 for Pilbara Iron, and 5128 for BHP Billiton).

However, the task of establishing the size of mining-related workforces at any one time is contentious. For one thing, workforce data inevitably contain a Perth-based component, while a distinction exists between direct company employment and workers engaged by contractors. Quite where the boundaries of this outer group of workers should be drawn is a moot point. For example, in 2003, the workforce employed directly by BHP Billiton in its Pilbara operations amounted to 2093 people, whereas a total of 4109 workers were employed by contractors engaged by BHP Billiton (BHP Billiton Iron Ore 2003: 9). Given the inherent market uncertainty in both the timing and scale of mining and related activity, both the balance and size of these relative numbers are subject to volatility. Notwithstanding these caveats, the ATAL unit of Pilbara Iron maintains a growing database of Indigenous and total employment as part of the company's Indigenous Employment Strategy and these are used extensively here to explore company aspirations for engaging Indigenous labour against regional demographic and socioeconomic trends.

Despite long-standing (and often independent) participation in mining employment in the Pilbara extending back to the early twentieth century (Holcombe 2004), Indigenous people have been only marginally engaged by the industry until very recently. For much of the period since the 1960s, the emphasis on rapid resource development in the Pilbara resulted in less concern for the welfare of Indigenous people, including in their employment (Cousins & Nieuwenhuysen 1984: 131–40). Not surprisingly, as recently as 1996, Indigenous people comprised barely 2 per cent of all workers in the Pilbara mining industry, and mining comprised only 10 per cent of all employment among Indigenous people in contrast to its dominance in the region among the non-Indigenous workforce. Largely in response to such disengagement, ATAL was formed in 1992 by Hamersley Iron with the aim of assisting Indigenous people to participate in and benefit from the company's operations. Training and employment opportunities have formed a key part of this strategy while, more recently, the establishment of Land Use Agreements with various Pilbara Aboriginal groups has increased and formalised the company's obligations in terms of heritage activities, thus widening the scope for engaging local Indigenous people.

Accordingly, Pilbara Iron has successively raised its Indigenous employment targets with a view to aligning these with the Indigenous share of regional population. As of April 2005, Pilbara Iron had 97 Indigenous staff in direct employment or training with the company, with an additional 63 employed by other companies who contract to Pilbara Iron. This represented 4.5 per cent out of a total workforce of 3555. Presently, the medium-term company aim is to ensure that the Indigenous share of its total workforce reaches a minimum of 15 per cent. Existing plans to attain this target, as outlined by ATAL under its current Indigenous Employment Strategy, extend to 2010. By this time it is expected that 12.3 per cent of the Pilbara Iron workforce will be Indigenous.

Continuing on this planned trajectory, a 15 per cent share is expected to be achieved by 2013.

Between 1992 and 2004, a total of 131 Indigenous trainees had passed through ATAL training programs. Of these, the majority (101) graduated while just under 10 per cent withdrew before completion. A total of 17 individuals were still in training as of August 2004. Of interest, in terms of Pilbara Iron's employment targets, is the fact that fully one-fifth of ATAL graduates are now employed by non-Rio Tinto companies, underlining the fact that companies compete for labour from the same common regional pool. Also of interest is the fact that very few graduates (just nine) were unemployed. Recruitment to the program has been largely, though not exclusively, from local language groups as indicated in Table 3.9 with around one-third drawn from Yindjibarndi and Banyjima.

Table 3.9. Group affiliation of ATAL graduates, 1992–2004

Group	No.	%
Yindjibarndi	23	17.6
Banyjima	22	16.8
Nyamal	12	9.2
Kurrama	11	8.4
Ngarluma	11	8.4
Yinhawangka	8	6.1
Thalanyji	5	3.8
Thursday Islander	4	3.1
Kariyarra	2	1.5
Pinikura	1	0.8
Jaburara	1	0.8
Niabali	1	0.8
Other groups	30	22.9
Total	131	100.0

Source: ATAL, Dampier.

Employment with Pilbara Iron in mid 2004 at the commencement of the current (2005) Employment and Training Strategy is outlined in Table 3.10. This shows the distribution of the Pilbara Iron and Robe workforce at that time by company worksite, including the Indigenous component and spread of Indigenous occupations. Overall, the Indigenous workforce was just 3.8 per cent of the total, but despite this small proportion the Indigenous workforce was spread across most worksites, though predominantly at Pilbara Iron sites. The majority were engaged at Pilbara Iron Dampier and Pilbara Iron Tom Price. As for occupations, the majority of Indigenous workers (65%) were employed as plant operators, with only one in a managerial role.

Table 3.10. Total and Indigenous workers at Pilbara Iron/Robe worksites, 2004

Worksite	Total workforce	Indigenous workforce	Indigenous occupation	No.
Robe Cape Lambert	342	18	Plant operator	7
			Liaison/Community	2
			Trainee Car Examiner	3
			Apprentice/Business	6
Robe Pannawonica	237	1	Business	1
Robe West Angelas	262	7	Plant Operator	7
Robe Perth	78	0	N/a	0
Pilbara Iron Dampier	710	35	Trainee Plant Operator	11
			Plant Operator	5
			Office Based	3
			Apprentice/Business	15
			Superintendent	1
			Trainee Operator	9
Pilbara Iron Tom Price	674	28	Plant Operator	11
			Liaison/Community	1
			Lab Assistant	1
			Apprentice/Business	7
Pilbara Iron Marandoo	121	6	Plant Operator	6
			Trainee Plant Operator	1
Pilbara Iron Paraburdoo	451	6	Plant Operator	3
			Manager	1
Pilbara Iron Channar		1	Plant Operator	1
Pilbara Iron Yandicoogina	162	5	Plant Operator	5
Pilbara Iron Brockman	128	0	N/a	
Pilbara Iron Perth	249	1	Office Based	1
Ieramugadu Crew	4	4	Gardener	4
Bridda Crew	14	14	Plant Operator	12
			Trainee Plant Operator	2
Other Companies Contracting to PI		6	Plant Operator	4
			Tradesman	3
Dampier Salt Pty Ltd		3	Plant Operator	3
Totals	3531	135		135

Source: ATAL, Dampier.

In addition, there are part-time work activities associated with site heritage clearance. Aside from working group and monitoring-liaison meetings paid for via the Pilbara Native Title Service, Pilbara Iron employs casual Aboriginal consultants for archaeological and ethnographic work. In 2004, Aboriginal heritage survey work involved 170 individuals (mostly males) from a variety of Pilbara Native Title Claimant groups and language groups including Gobawarrah Minduarra Yinhawanga (GMY), Wong-Goo-tt-oo, Yapurarra Martuthunira, Kurrama, Ngarluma, Yindjibarndi, as well as the Innawonga Banyjima Nyiyaparli claimant group. Within the framework of the Indigenous Employment Strategy, this survey work is seen as providing ongoing opportunities for part-time work on country, amounting incrementally to the equivalent of an additional eight full-time positions each year. In 2004, such work accounted for a total of 1300 consultant days. However, such work is not regular and represents more of a windfall activity spread across select groups of people, although demand for labour is directly tied to the amount of land disturbed and so is likely to keep pace with the expansion of exploration and mining.

Pilbara Iron (and other companies) also contribute to Indigenous employment via support for contracting businesses, either directly through Indigenous contracting businesses, or by stipulating the use of Indigenous labour quotas for other contractors. For example, the 1997 Yandi Land Use Agreement opened the way for the establishment of joint venture businesses between the Gumala Aboriginal Corporation and Hamersley Iron. Today, Gumala Enterprises manages two such businesses – Gumala Contracting, an earthworks business which contracts to Pilbara Iron and other industry and government organisations, and Gumala Eurest Support Services, which provides cleaning and gardening services to Pilbara Iron sites and services the Savannah Campgrounds facility at Karijini National Park. Presently, the contracting arm employs 13 people of whom seven are Indigenous. It is anticipated that by the end of 2005 contracting will employ 20 people of whom 10 to 15 will be Indigenous.

In addition to this, Brida Contracting was established in a collaboration between ATAL and the Roebourne community. Brida is fully owned by the Ngarliyarndu Bindirri Aboriginal Corporation in Roebourne and presently employs 27 people on earthworks and camp landscaping, 23 of whom are Indigenous. Current work is associated with Pilbara Iron's Dampier port expansion, which was due to conclude in August 2005. However, further work is likely to come via other Pilbara Iron and Woodside infrastructural expansions, and company estimates point to the prospect of employing 70 to 90 Indigenous people by the end of 2006. Other enterprises include the Wanu Wanu and Ngurra Wangkamagayi cross-cultural training businesses in Tom Price and Roebourne respectively. These companies provide an important input to capacity building on the non-Indigenous side of the workforce in order to assist in making Pilbara Iron operations more amenable to Indigenous workers (see Interview segment 3, p. 57; Interview segment 8, p. 59). Recently, a business alliance has also been formed between representatives of the Eastern Guruma group, contracting company Civil Road and Rail, and Pilbara Iron, associated with contracting out services to Pilbara Iron's rail duplication project (ATAL 2005).

The major Indigenous enterprise in the region is Ngarda Civil and Mining. In 2005, this company employed 170 personnel with Indigenous workers accounting for as many as 140 of these in line with the company aspiration of maintaining a minimum Indigenous share of its total workforce of 85 per cent. Ngarda provides contracting services to the mining and construction industries including contract mining, earthworks, road and rail construction and maintenance, mine site support services, haul road maintenance, crushing and screening operations, rehabilitation and environmental services. In 2004 a total of 81 Indigenous workers were evenly divided between Pilbara Iron (Pannawonica) and BHP Billiton (Finucane) worksites. Other significant enterprises include Indigenous Mining Services Pty Ltd in South Hedland which had 21 registered workers in 2004, while Woodside and Brambles Industrial Services also have agreements with Nyamal Crane Hire and the Ngarluma and Yindjibarndi Foundation at Roebourne for the provision of employment and training opportunities via the LNG4 expansion project. While precise numbers employed in these ventures at any one time are unclear, and while they may change according to normal business fluctuations, it appears that collectively the Indigenous business enterprise sector is a significant contributor to the regional Indigenous employment profile, possibly accounting for somewhere in the region of 200 workers, or around 15 per cent of Indigenous employment in mainstream work.

To appreciate the dynamics of the ATAL strategy in raising employment levels across all of these activities, it should be understood that its employment target is built around the notion of a potential Indigenous labour supply that is resident within the region but has varying human capital capacities and employment aspirations. For this population, the routes into Pilbara Iron employment are varied and include a combination of on-the-job apprenticeships and traineeships, as well as direct employment in mining operations and Indigenous enterprises. These opportunities are backed up by education and work-ready programs as well as pre-employment training. The Pilbara Iron Aboriginal Employment Strategy is therefore constructed around four strategic action areas:

- Capacity building: this includes education initiatives, scholarships, pre-employment training, fitness-for-work programs, and programs directed at reducing the impact of alcohol and other drugs.
- Training and direct employment: this includes traineeships, apprenticeships, earthworks, clerical training, and direct employment strategies.
- Improving retention: these are support strategies to assist in holding on to workers once employed, they include cross-cultural training across the workforce.
- Business development: this is a long-standing strategy to work with Indigenous individuals and groups to develop viable business enterprises, mostly contracting businesses.

The manner in which these strategies seek to direct Indigenous labour into Pilbara Iron operations is summarised graphically in Figure 3.6. As can be seen, most routes into Pilbara Iron employment involve some form of up-skilling or remedial input from the company. Aside from the capacity building activities that are directed towards increasing the potential stock of employable labour within the region, company inputs are mainly via (STEP) apprenticeships and various on-the-job traineeships. The other main flow into Pilbara Iron operations involves the interaction with Indigenous contracting businesses, as well as other businesses that seek to engage Indigenous labour. Interestingly, these also generate their own inter-business flows, although no data on these exist. What this diagram does not show, of course, is the leakage from this single company system to and from other components of the regional labour market – for example, to and from other mining and private sector companies, to and from public sector agencies, and to and from Indigenous organisations (including CDEP schemes). In the latter case, issues arise with regard to the potential impact of the drive to meet employment targets on source communities with limited skilled personnel (see Interview segment 5, p. 58; Interview segment 12, p. 61).

Figure 3.6. Indigenous employment routes into Pilbara Iron operations

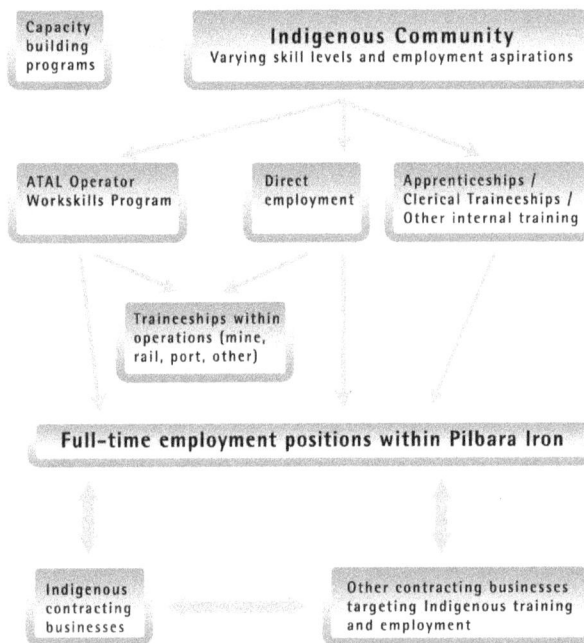

Source: ATAL 2005.

In estimating the growth of Pilbara Iron's Indigenous workforce towards the achievement of the company target of a 15 per cent Indigenous share of total workforce, ATAL uses these component routes into employment for its projections. Thus, commencing with 160 Indigenous employees at the end of 2004 (including those with contractors), the current ATAL schedule aims to see this number added to by 20 workskills employees, 5 direct employees, 20 apprentices in training, 18 apprentices into employment, 15 port/mine trainees into permanent positions, and 8 full-time equivalent heritage positions, less a

15 per cent turnover, to produce an Indigenous workforce by the end of 2005 of 209. If this were to be achieved, it would represent 6 per cent of the total Pilbara Iron workforce. The current plan would see this number grow in similar fashion to 430 (12.3% of the total workforce) by 2010. Against this trajectory, the ultimate goal of 15 per cent of the workforce would require a total Indigenous workforce of 525 (an extra 365 positions compared to the 2004 level), and would probably be achieved by 2013. If we hold the current proportion of male to total Indigenous employees constant (at 85%), this would imply an additional 310 male and 55 female workers by that date.

In order to benchmark these targeted outcomes against existing regional employment levels, Figs. 3.7 and 3.8 outline the estimated numbers of Indigenous males and females aged 15–54 in 2006 (based on the projection in chapter 2) according to their different labour force status categories (based on 2001 Census-based rates and 2005 CDEP participant numbers). Thus, out of a total of 2092 males in the prime working age group, an estimated 774 will be in mainstream work in 2006, and 445 in CDEP. Based on census rates, 200 would be unemployed, while those not in the labour force (673) would be almost as many as those gainfully employed. As noted earlier, confusion surrounds the true level of unemployment (and therefore the size of the labour force) owing to the large number of Indigenous adults on Newstart Allowance.[2]

If we compare these levels of 'current' (2006) employment, we can see that the Pilbara Iron target for male employees amounts to more than half of the 'current' number that is 'presently' employed in full-time mainstream work in the Pilbara region as a whole. As for females, the target amounts to around one-third of the 'current' mainstream full-time employment level. Set against these levels, the targets set by Pilbara Iron represent potentially substantial impacts on regional employment levels. However, Pilbara Iron is not the only corporate seeking to meet Indigenous employment targets – BHP Billiton, for one, has a 12 per cent target by 2010 which would translate into something like 300 Indigenous workers, or an increase of two-thirds over existing levels (as at February 2005 approximately 8–10% of BHP Billiton's Iron Ore's operational workforce was Indigenous). In effect, the various plans for Indigenous engagement in the region, from the resources sector alone, appear to have the capacity to virtually double the size of the full-time Indigenous workforce in the mainstream labour market, at least in theory.

---

[2]For confidentiality reasons, a breakdown of these Newstart customers by age and sex was not available from Centrelink.

Figure 3.7. Estimated numbers of Indigenous males aged 15–54 by labour force status: Pilbara SD, 2006

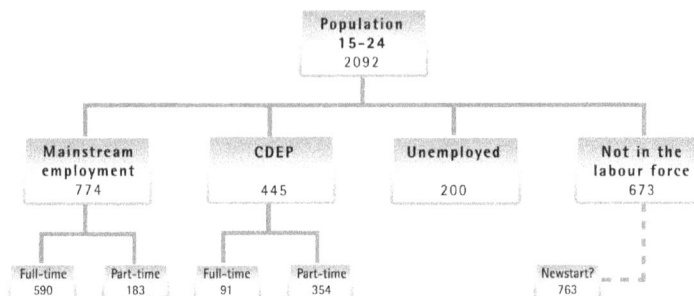

Note: CDEP figures based on 2005 participant numbers. Newstart figure is for all such payments (no sex breakdown available).

Figure 3.8. Estimated numbers of Indigenous females aged 15–54 by labour force status: Plibara SD, 2006

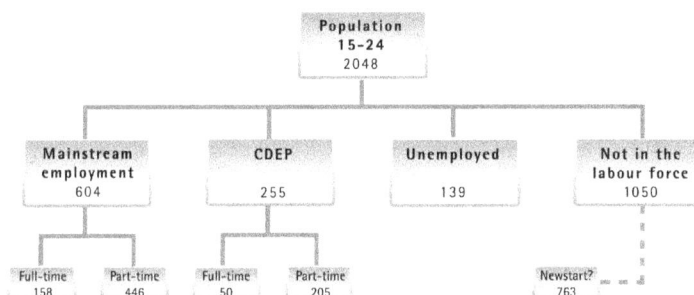

Note: CDEP figures based on 2005 participant numbers. Newstart figure is for all such payments (no sex breakdown available).

With this observation in mind, a number of questions arise as to feasibility. First of all, if we examine Figs. 3.7 and 3.8 and consider the potential sources of this labour, we can see that targeted expansion on this scale can only be achieved by drawing from the ranks of those in part-time mainstream employment and CDEP (many of whom would have some work experience), but also from the pool of those unemployed. However, this assumes that all such labour is capable of being directed to the mining sector, or even wants to be, which, of course, is unlikely to be the case. As indicated in Table 3.4, only 72 per cent of the non-CDEP Indigenous workforce is in the private sector, with Indigenous people more likely than the rest of the Pilbara labour force to be in public sector jobs, notably in community services, education and health (Figs. 3.2 and 3.3).

While some of this sectoral distribution reflects the relative lack of industry-based skills and qualifications among Indigenous adults, some of it no doubt also reflects choice, particularly for work in the community services sector. Unlike the non-Indigenous workforce that the mining industry can (and does) augment by recruiting from outside of the region, the basic premise of Indigenous workforce planning under existing corporate social responsibility goals, and in line with regional Indigenous aspirations and agreement-making, is that labour be sourced primarily from within the Pilbara. The extent to which this is presently the case is ultimately unknown since a number of Indigenous workers do not reveal their group affiliation in Pilbara Iron records. Nonetheless, according to 2004 data for the Pilbara Iron Indigenous workforce, it would appear that only 62 per cent of those who stated a group affiliation are locally-sourced, as indicated in Table 3.11. Of course, if we were to include those engaged in heritage survey work, then the local component of the workforce would rise.

Table 3.11. Group affiliation of Indigenous workers at Pilbara Iron, 2004

	No. identifying	% of those identifying
Yindjibarndi	17	14.9
Nyamal	15	13.2
Thursday Island	13	11.4
Banyjima	12	10.5
Noongar	11	9.6
Kurrama	9	7.9
Ngarluma	9	7.9
Other groups[a]	28	24.6
Not stated	21	N/a
	135	

[a]Includes a wide range of individuals including from the within the Pilbara (Yinhawangka and Thalanyji) as well as Bunuba people, along with people from the Carnarvon area, Murchison area, Broome Area, Kalbarri Area, and Yorta Yorta people.
Source: ATAL, Dampier.

However, even if Indigenous labour supply were spatially confined to the Pilbara, it is still expanding since the local Indigenous population of working age continues to grow at 1.6 per cent per annum. At the same time, this labour is sought by a variety of potential employers, many of whom (particularly in the mining industry), are in competition for the same distinct labour pool owing to the specialist skills involved (as noted earlier, up to one-fifth of graduates from ATAL are employed by non-Rio Tinto operations). There is also the question of population distribution within the Pilbara to consider. If the population numbers in Table 2.3 are read in combination with Fig. 1.1, it can be seen that almost two-thirds of the Indigenous population (64%) is located in and around coastal towns, whereas a large share of the jobs associated directly with mining are at distant inland sites far from where most people live. Even for those resident inland, unless they are located within one of the company townships, physical access to mine work becomes an issue. For the most part then, mining employment for Indigenous people requires some form of FIFO, or at the very least DIDO arrangement, requiring prolonged absences from a home base, something that can be viewed negatively in a society where responsibilities to family and kin are important considerations (see Interview segment 5, p. 58).

All of this aside, employment in mining, or in any other form of mainstream workforce engagement, may simply be a lower priority than the pursuit of more customary culturally-based activities (see Interview segment 5, p. 58; Interview segment 33, p. 94). However, there is also the much wider and structurally intransigent issue of the behavioural disengagement of many younger Indigenous people from mainstream (and customary) institutions that emerges from many of the interviews and that places substantial emphasis on the need for capacity building programs to increase their range of options (see ; Interview segment 1, p. 57; Interview segment 9, p. 59; Interview segment 10, p. 60; Interview segment 15, p. 61, Interview segment 27, p. 93; Interview segment 34, p. 94).

There is no question that ATAL's activities, and the general climate of economic growth in the Pilbara as a whole, has raised expectations among Indigenous residents about their prospects for enhanced participation (see Interview segment 1, p. 57; Interview segment 2, p. 57; Interview segment 11, p. 60; Interview segment 14, p. 61; Interview segment 16, p. 62; Interview segment 23, p. 75) . However, the combination of demand- and supply-side issues outlined above raises a number of questions that go to the matter of whether or not these expectations are likely to be fulfilled. First of all, how many additional Indigenous jobs will be required over the coming years in order to achieve particular outcomes in terms of regional labour force status given the fact of high population growth? Second, what is the potential of Pilbara Iron's Indigenous Employment Strategy to make inroads into these requirements? Third, what social, behavioural, and human capital factors might impinge on the supply-side to influence the achievement of these potential outcomes? The first two questions can be addressed immediately. The last question provides the stimulus for the remainder of this monograph.

Future employment needs

The lack of noticeable improvement in the labour force status of Indigenous people in the Pilbara over the last four decades is a function of their sustained inability to increase participation in the mainstream labour market in line with their rate of population growth. In effect, such participation has gone backwards as a proportion, with expansion in work opportunities for Indigenous people occurring instead largely via the CDEP scheme. At the same time, the thrust of current government policy aimed at reducing welfare dependence and raising economic status is, of necessity, towards increasing mainstream employment. In a region such as the Pilbara, this would have to place primary emphasis on the private sector.

As for labour demand in the region, the December 2004 version of the biannual labour market forecasts produced by the Centre of Policy Studies at Monash University indicate an increase of 1469 persons of prime working age in jobs in the Pilbara between 2004–05 and 2011–12, with two-thirds of these for males. As noted earlier, these estimates are likely to be conservative given their lack of local economic intelligence and input. For example, if we take resource projects alone in the Pilbara that are either underway or planned as at the beginning of 2005, these will require an estimated temporary construction workforce of 11 152 and permanent employment for an additional 2165 (Government of Western Australia 2005: 25), to say nothing of the employment multipliers generated by minerals industry expansion in Western Australia in areas such as services (Clements & Johnson 1999). Nonetheless, the Monash estimates provide a clear indication of sustained growth in employment demand in the years ahead, and certainly over the period for which Indigenous population quotas have been projected. At the same time, the Indigenous population of working age is also set to rise at around 1.6 per cent per annum. What then is the scale of the task ahead in terms of creating new job opportunities for Indigenous people if the aim of policy is to improve their overall socio-economic circumstances beyond the current low level? To establish this, we can use the projection of the future size of the Indigenous working-age population and consider this against expected growth in employment.

Overall, by 2016, the Indigenous population of working age is projected to increase by 17 per cent from 2006 to reach a population of 5555 – an increase of just under 800 persons. As shown by the percentage change for different age groups, much of this growth will occur in the older ages over 50 years. Realistically though, it is those in the age range 15–49 who are likely to be targeted for emerging opportunities in the regional labour market, and this age group is set to increase by a total of 477 persons over the 10 years to 2016.

Against the background of these population projections, Table 3.12 explores three future employment scenarios. The first considers the number of jobs that would be required by 2016 if the 2001 Indigenous employment to population ratio were to remain unchanged at 42.5 per cent (inclusive of CDEP). The answer is 2361, or an additional 553. This is in order of magnitude to the number of new jobs sought by Pilbara Iron and BHP Billiton combined in their Indigenous employment strategies. The second scenario considers the extra jobs required to maintain the reported mainstream employment population ratio of 30.2 per cent. This would require fewer jobs at 1678, an increase of 394 compared to 2001. If, however, the aim were to achieve parity with the regional non-Indigenous employment to population ratio of 80.3 per cent, then the number of Indigenous people in work in the Pilbara would need to more than double to 4460, requiring an additional 2652 jobs.

Table 3.12. Extra Indigenous jobs required in the Pilbara between 2001 and 2016 against select employment rate targets

Employment/population ratio in 2001	Base employment 2001	Total jobs required by 2016	Extra jobs required by 2016
42.5[a]	1808	2361[b]	553
30.2[c]	1284	1678	394
80.3[d]	1808	4460	2652

[a]The 2001 census-derived Indigenous employment/population ratio inclusive of CDEP.

[b]The 2001 census-derived Indigenous employment/population ratio exclusive of CDEP.

[c]The non-Indigenous census-derived employment/population ratio in 2001.

[d]Based on projection of working-age population to 2016 (5555).

What these projections suggest, is that the combined Indigenous employment targets currently set by Pilbara Iron and BHP Billiton, if achieved, will be more than sufficient to raise the Indigenous employment rate in mainstream work (exclusive of CDEP) to 33 per cent, will be just sufficient to maintain the 2001 employment rate (inclusive of CDEP), but (alone) will be far from sufficient to begin to close the gap with the non-Indigenous rate. In effect, because of population growth, the combined efforts of Pilbara Iron and BHP Billiton, substantial as they are in terms of potential increase in numbers employed, will only manage to keep pace with the extra numbers entering the working-age group. Thus, in terms of improving Indigenous labour force status to anything even approaching the norm for non-Indigenous residents of the Pilbara, this task is way beyond any impact that could emanate from planned mining employment.

Of course, the mining sector is not the only employer of Indigenous labour, now or into the future. Overall, Indigenous employment growth in the region between 1996 and 2001 was relatively high (5.6% per annum including CDEP), and this during a period of overall decline in labour demand in the Pilbara as indicated by a reduction in the overall workforce from 21 466 in 1996 to 21 058 in 2001. This reflects the segmented nature of the Australian labour market in terms of Indigenous engagement in so far as Indigenous people disproportionately occupy particular sectors (public/community/CDEP) and have been the focus of affirmative action programs that can produce counter-cyclic employment trends. On this basis, if we assume (perhaps optimistically) a continuation of this overall employment growth rate for Indigenous people then the overall prospects of significantly raising the regional employment rate appear promising given the solid base input from the mining sector.

Indigenous perspectives

Interview segment 1

We are trying to get our kids into those mining training areas to get ready for jobs coming up. Some of them don't have the education, or too much drugs or alcohol. We try to counsel them and say you got to get into these jobs, you need to prove yourselves and stop all this other nonsense. Lot of people don't get jobs because they right without, particularly with access to drugs and alcohol these days, they don't want to work. That's a big problem.

Interview segment 2

All those new jobs coming up, well they'll be fly-in-fly-out. Yep. They don't give Aboriginal people the opportunity. You might get some good educated ones put in their resume and everything like that, but they don't even get an interview. And that's letting them down, and they think, 'forget it, I've had a go, my application was good enough to have an interview and I don't get one'. They just feel let down and that's when they let their self esteem go down, finish. And they don't worry about applying again. The best thing to do with Aboriginal people is the hands on, get em in there on job training. But they don't even give them that opportunity.

They reckon they got jobs after that training for us locals, and they do that training, you know, the local people, they stick to that training, and they think to themselves, 'I done all this training I might have a full time job then', but they aren't guaranteed a full time job which is very very sad, it's rough. It's all fly out and fly back, fly out and fly back! We don't want em' them peoples like that, and you know they say its there for the local people.

Interview segment 3

Some of those white fellas in mining are racist too you know, they say, 'ah bugger Aborigines they got no brains at all, they won't hang in there long', and all those sort of things, you know. And our mob need the support when they get into those sorts of positions you know, encourage them more. If they can't pick up something well at least like supervisor go there and have a talk to them nice way, like not just tell 'em to just get on with that job, Aboriginals got a different feeling, you gotta have a really good person to talk to those fellas you know, the white bloke gotta come up and help those young fellas, and you gotta have another Aboriginal person, older one, old enough to tell them 'no don't do that' and help 'em, learn 'em you know?

Interview segment 4

We got an art and craft program funded by CDEP and sewing but the main thing we need is this office here, the community office, so I'm doing work there and underpaid. The community doesn't pay me for that. Working in the office is part of my CDEP hours. The government taken away ATSIC and they put in ICC, Indigenous Consultant Committee, an organisation that has been established by the government. Remember ATSIC and ATSIS? When the government took away ATSIC they set up ICC, well there's no more ATSIC, and ICC investigated those two organisations. ICC, well it's still not what we want from them, what the community needs, it's what they think what's best for the community, for the people, see? But we want a full-time administrator in the office, a coordinator. We need someone who is tough and tell people what they need to do, and they gotta be paid to do it. But that ICC, they came here, but they ignored that.

Interview segment 5

Life is a bit better out here because of mining and those agreements. But my thing is my own kids, we're not pushing them into what we want them to be. It's up to them as individuals. I believe that fair enough, they can go and work in the mine, but they will be men and will have kids of their own, and they need to be there for their own kids to learn and teach them their culture. Because its about carrying on the traditional cultural ways teaching knowledge skills, and the country itself, all those kinds of things, the trees the language, going to ceremony, going out on country. My kid's father is teaching our kids. His grandmothers and grandfathers, they passed on all the knowledge to him, making him understand who he is, he hasn't missed out on anything, he's got it all and he knows what his role is as a cultural man and in our cultural life. But some of those mining men aren't there for all that and that's no good.

Interview segment 6

Government has created a big problem by creating dependency. There's plenty of work around in these parts. When you look back at the 60s and 70s there was plenty of work for our people, yet we were kept out of that main workforce. My parents and my wife's parents worked, I suppose they were a little bit more education and others were a little bit shy because they didn't have that education, so a lot of them were kept out of the workforce. That's where it started, when Aboriginals one side and Whitefellas one side. It was all that old policy that contributed to what's going on now. The half-caste people were getting jobs, where the pure bloods weren't, so if you had a lighter skin then you had a chance, whereas if you had a dark skin you didn't have a chance. So it was still a policy of assimilation and the half-caste kids were taken away. It's still a bit like that too you know.

Interview segment 7

Look at Tom Price, why aren't they building that into a town instead of making it all for fly-in-fly-out, it's all 12 hour shifts and there's no balance in the work.

They might be doing two weeks on and one week off, but that time is also splitting their family up, that time away. I used to work the three eights you know, eight-hour shifts. It was balance, and I was living in Wickham at the time. The town was prospering in the social life because they always had sports on, and everybody was socialising, there was a good atmosphere in the town, and Tom Price was the same, a lot of sports activities football and basketball, leisure time and speedway. But now you go there, you don't see that anymore, it's all one way. It's all work work.

Interview segment 8

I've worked in mining and I know the mining world is the toughest of all. What a lot of our children don't have is that self-esteem, and the confidence to work with a non-Indigenous person, and to take all the crap that comes out of their mouth, and not take it in personally. You know what working people can be like! That is sad and that's that education part. But some of our young people are strong enough to have that confidence and say, 'no, I'm going to go to college and learn myself to do this, and learn a bit more'. I wish there was more of them.

It's hard for women to work too, you can't afford child care unless you are a high income earner, yeah that's an issue, and it puts a lot of pressure on old people to look after kids, and that family stuff comes into play, but it's the age barrier that's the biggest thing of all. You can't expect your grandmother to look after your kids when she's as blind as a bat and can hardly even walk!

Interview segment 9

My concern is more to do with the numbers that are not preparing themselves for all this employment that is coming up. Its not the responsibility and the problem of the mining companies, it's the community. Why aren't we preparing our young people to move into better jobs? Are there things in Roebourne that can be tailored so that young people can step through a career path? There are a lot of kids that have not been given any direction in terms of love, care, parenting and education. They don't even know the importance of attending school or behaving well, and for instance they say things like, 'why should I be in bed by 10pm?', all of those factors make a person think in terms of whether they are going to go on in life or not. If you got a lifestyle of staying up late, and no one stops you from doing it, I can see that they have a problem with boundaries not being set. There is no discipline in terms of commitments, I'm not talking about punishment, I'm talking about the person committing him or herself to a direction, it's personal discipline. Because the parents aren't around young people miss out the love and the care of a family home, and that tells me that as soon as that is not there, that child no longer knows what will be expected of them as parents when they get older. So what that child will pass on to their own kids is what they learnt off the streets. This all comes from a combination of alcohol and drugs and just plain poverty.

Interview segment 10

Not everyone is a drinker, but in families where there is drinking there is a likelihood of separation, or it has already happened. In such cases kids usually end up living with grandparents or extended family, or they just move around from household to household with mates. Now you look at all of that, what is the chance of that guy trying to get to college? They might get to 16 or 17 and think, 'I got to go to college to try and make something of my life', and I got to give credit to some of those young fellas who make that decision and get so far, and yeah they might fail, but they should be given credit because a lot of times they have tried to come from having absolutely nothing. There's no pick up programs for kids like that. The Woodside employment and training program picks up a little bit. The education program (Gumala Mirnuwarni Education project) only picks the cream of the crop, which isn't fair on the ones that are never going to make it. I like seeing those that are less equipped in life being picked up and given the opportunity. That doesn't happen much in Roebourne. Then you have the assessments with Hamersley Iron or Woodside or another employer group, who wants to take people on. You'll get 20 people there, which is normal; I don't say that they shouldn't get every one in. But you know already that out of that 20 you will only get 5 because that other 15 aren't going to be able to answer the first lot of questions. So there is not a development program for up-skilling. There should be a program teaching 'why is it important for me to get to work on time, why is it important for me to go every day to work, why is it important for me not to drink heavily during my work period', all those kind of things. The employers might look at that and think that's molly-coddling, but we are talking about people who have no idea or don't even want to.

Interview segment 11

We got a lot of kids out there too. A couple are working age, one teenager, the one who went through business. He just started on CDEP, I got him working out there keeping him out of mischief. Work they do is clean the toilets and water the trees. We are trying to get our kids into those mining training areas ready for that kind of thing. That ATAL, getting trainees through there to get ready for jobs coming up. Problem we have is that our people are lower in numbers than other Aboriginal people getting into the program, people from other areas. I don't know why, maybe Banyjima people don't have the education or too much drugs or alcohol or bloomin' thing.

Interview segment 12

We got to target the young to do it because this age group now, the ones who are in the 18s are getting lost and dependent. On the other side of the coin, the good go, the educated who want to do something they can go on and do things. They go in the mining companies, or wherever. Then we are left with the ones who aren't fortunate. But where it starts to fall down is that we need workers in the community, and we haven't got the real skilled people in the community. They spend too many hours in the mining industry and have no time for family. We have lost kids here now as a result of all that.

Interview segment 13

When they first started making the railway line something went wrong with the government and they didn't want Aboriginal people to work on it, but they brought other people from the Torres Strait Islands, to build the railway line. They stopped hiring Aboriginal people. Something went really wrong back then. I'm a machine operator. We used to be able to get a job anywhere, but now you got to be able to go through the school. You got to know computers too. I put in for a job with the miners and they asked me if I could read and write. They also told me I had to learn how to use a computer. Well blackfellas don't need a computer, they got a computer in their heads! I know all about engines, how to fix a car, how to drive truck, plant equipment all that. Well we can't get a job in mining so we started our own business.

Interview segment 14

My understanding ATAL belongs to us, we have an agreement with Hamersley Iron if we really want to get down to the nitty gritty of it, but they also have a commitment to us as we have a commitment to them and that is to ensure that employment is there and waiting for these people at the end of their training. I mean I look at their fly-in-fly-out and these people are only here for x amount of years and then they're gone, what commitment do they give, compared to the ones that come from here? There's that problem, mmm.

Interview segment 15

There's a lot of services in Roebourne, lots of organisations, and visits from training groups, its been available for quite a number of years, but the problem is if a person doesn't want to give it a go then there's not much that can be done. When you have a town that is welfare based like Roebourne, the thinking generally is along the lines of, 'I don't want to do any training coz I'm getting my fortnightly pay', and it doesn't do them any good because they don't look at their future. A lot of people look at it like, 'I'm only going to live in Roebourne, and there's nothing here, so why should I go to train to be something'. There's no outside thinking, I don't have any answers for that, but there should be ways and means to give those young people encouragement to work further afield before they return to this town. They will bring back skills, and be more positive about the future. If you go away from a small town, or a community, you can see a lot better. You can see the needs but you can also see other things, a bit like sitting in the back row and observing.

Interview segment 16

We got a few people working now. Hamersley put out a lot of training thing now. What they tell me, they go and do all the training at Pundulmarra College, but then they say, 'where we gonna go then, no jobs?' I been talking to mining companies, 'if you are going to give them a training, then you should give them a job too'. That word has gone around now, so nobody want to do the training, they think, 'well what's the point?' I think the problem is that they are hiring people from elsewhere with better qualifications than we can get around here. Sometimes they do ten month training but then get no job, but they say there are biggest mob of jobs, but they aren't training people to the right standard if they can't get a job at all. Also they are using government money to provide that training for mining.

4. Income status

Indigenous people in the Pilbara have a number of potential sources of cash income. These range from wage labour in mining and other mainstream forms of work, to participation in a CDEP scheme, unemployment benefit and other payments from Centrelink, royalty or other agreed payments to traditional land owners, and private income from the sale of art works, crafts and other products. Set against these, of course, there are routine deductions from income at source, such as those for house rent and power charges.

Accurate data on income levels, and employment and non-employment sources of income, are notoriously difficult to obtain due to a variety of conceptual problems. For one thing, most measures of income refer to a period of time, such as annual or weekly income, whereas the flow of income to individuals and households within the region is often intermittent. Census data, for example, are collected for all sources of income in respect of a 'usual week' and then rounded up to annual income. What might constitute 'usual weekly' income in many Indigenous households is difficult to determine. On the credit side, there is the likelihood of intermittent employment and windfall gains from sources such as gambling, cash loans, and royalty payments. This sort of income combines with debits (for example, due to loss of employment and cash transfers to others), to create a highly complex picture even over a short space of time, and one that census methods of data gathering are likely to misrepresent.

Even if adequate questions were asked regarding income, high levels of population mobility would make it difficult to establish a consistent set of income recipients over a period of time. This is further complicated by job mobility with individuals often employed on a casual or part-time basis and moving into and out of longer-term jobs. As for the circulation of cash between individuals and households, information on this is non-existent. Also lacking are data on expenditure, although a common pattern reported from elsewhere is one of cash feast and famine against a background of high costs for essentials such as food and transport (Taylor & Westbury 2000).

The most comprehensive source of income data for the region based on a consistent methodology is the census. It should be noted, however, that census data report income in categories, with the highest category left open-ended. Consequently, actual incomes have to be derived. In estimating total and mean incomes, the mid-point for each income category is used on the assumption that individuals are evenly distributed around this mid-point. The open-ended highest category is problematic, but it is arbitrarily assumed that the average income received by individuals in this category was one-and-a-half times the lower limit of the category (Treadgold 1988). Clearly, estimates of mean incomes will vary according to the upper level adopted.

Also, the gross income reported in the census is intended to include family allowances, pensions, unemployment benefits, student allowances, maintenance, superannuation, wages, salary, dividends, rents received, interest received, business or farm income, and worker's compensation received. Whether all such sources are reported is unknown. One distinct advantage of census data, however, is that it provides a means by which an estimate of dependence on income from welfare can be derived. This is done by cross-tabulating data on income with labour force status as a basis for distinguishing employment income from non-employment income, the latter being considered a proxy measure of welfare dependence.

Employment and non-employment income

The relative contribution made to total income from employment as opposed to from other sources is an important factor in terms of the economic situation facing Indigenous people in the Pilbara regional economy. Approximate parity between net incomes derived from social security and those derived from employment (after tax) is likely, unless there is sufficient participation in well-paying jobs. It is argued generally for Indigenous people that the gap between welfare and earned income is sufficiently low as to discourage job seeking (Hunter & Daly 1998). In the Pilbara, given the large and expanding number of relatively high paid positions in the mining industry, this is potentially less of an issue, with the challenge more about ensuring job readiness for the opportunities that are available.

Fig. 4.1 shows the relative personal income distribution for Indigenous and non-Indigenous adults enumerated in the Pilbara in 2001. Enumeration data are used here in order to include in the profile the incomes of temporary workers (mostly non-Indigenous) who are employed in the Pilbara but whose usual residence is elsewhere. Of particular note here are FIFO workers who earn their income in the Pilbara but who invariably spend it elsewhere (see Interview segment 20, p. 74). It is clear that the bulk of Indigenous people have incomes at the lower end of the distribution with a clustering around $199 per week ($10 400 per annum) and a steady falling off thereafter. More than three-quarters (78%) of Indigenous incomes are less than $500 per week ($25 968 per annum) which is the point at which the two distributions cross over. In contrast, 68 per cent of non-Indigenous incomes are above this level, and the non-Indigenous distribution rises at the higher end with more than one-third of incomes (36%) above $1000 per week.

Figure 4.1. Weekly personal income ($) distribution of Indigenous and non-Indigenous adults in the Pilbara, 2001

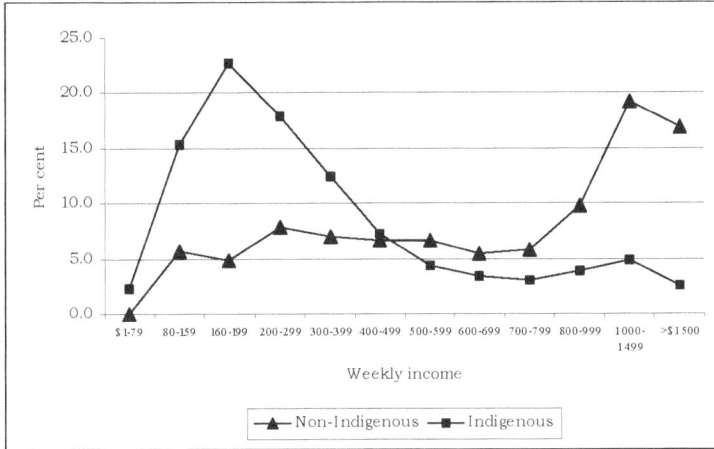

Source: ABS 2001 Census customised place of enumeration tables.

Indigenous household incomes also fall substantially behind those of non-Indigenous households (Fig. 4.2) with almost half of all Indigenous household incomes (47%) falling below $700 per week. By contrast, as much as 86 per cent of all non-Indigenous household incomes fall above this level, with 48 per cent in the top two income brackets over $1500 per week.

Figure 4.2. Indigenous and non-Indigenous weekly household income: Pilbara SD, 2001

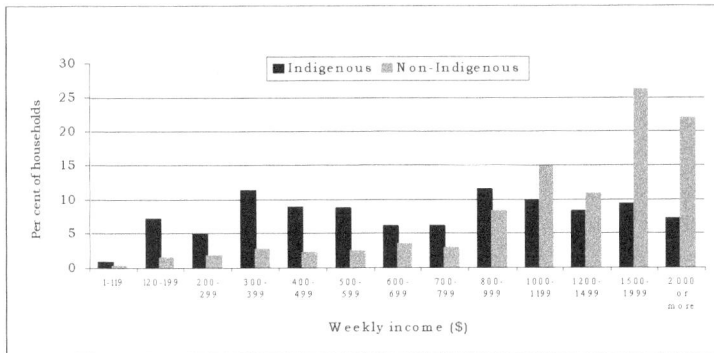

Source: ABS 2002b.

As with many social indicators, the Pilbara average masks considerable diversity. Table 4.1 shows the distribution of Indigenous and non-Indigenous median family and household incomes by IA, together with average household sizes. While the gap in relative incomes is everywhere substantial, this is exacerbated for Indigenous households by their larger household size and need to support a greater number of dependents on smaller incomes (see Interview segment

17, p. 73; Interview segment 18, p. 74; Interview segment 20, p. 74; Interview segment 23, p. 75). The difference is most acute in Port Hedland where Indigenous households have the added burden of competing for housing. Karratha stands out as having by far the highest Indigenous family and household incomes.

Table 4.1. Indigenous and non-Indigenous median family and household weekly incomes by IA: Pilbara SD, 2001

Indigenous Area	Indigenous family income ($)	Indigenous household income ($)	Mean Indigenous household size
Roebourne	700–799	700–799	4.2
Roebourne bal	500–599	700–799	3.9
Karratha	1000–1199	1000–1199	3.4
Ashburton	600–699	600–699	3.6
East Pilbara (W)	400–499	800–999	6.8
East Pilbara (E)	400–499	700–799	4.0
Jigalong	400–499	600–699	6.7
Marble Bar	500–599	500–599	3.8
Yandeyarra	500–599	700–799	4.3
Pt. H. (ex urban)	300–399	400–499	4.3
Port Hedland	700–799	600–699	3.4
Indigenous Area	**Non-Indigenous family income ($)**	**Non-Indigenous household income ($)**	**Mean non-Indigenous household size**
Roebourne	800–999	600–699	2.2
Roebourne bal	1500–1999	1200–1499	2.9
Karratha	1500–1999	1200–1499	2.9
Ashburton	1500–1999	1500–1999	2.8
East Pilbara (W)	600–699	600–699	2.5
East Pilbara (E)	1500–1999	1200–1499	2.7
Jigalong	n/a	1000–1199	1.6
Marble Bar	800–999	700–799	2.2
Yandeyarra	1500–1999	1000–1199	1.5
Pt. H. (ex urban)	800–999	800–999	2.8
Port Hedland	1500–1999	1200–1499	2.7

Source: ABS 2002b.

Table 4.2 shows these same data converted into annual average personal incomes according to the labour force status of Indigenous and non-Indigenous adults. For ease of comparison, the ratios of Indigenous to non-Indigenous incomes for each of these categories are shown in Table 4.3. Not surprisingly, for Indigenous people, employment in the mainstream labour market returns much higher personal income compared to work in CDEP schemes (see Interview segment 17, p. 73; Interview segment 18, p. 74; Interview segment 19, p. 74). However, Indigenous people in mainstream work still lag far behind their non-Indigenous counterparts, with the average Indigenous income from employment below two-thirds of the non-Indigenous level. Also of note is the fact that Indigenous income from non-employment (welfare) sources is substantially lower than the non-Indigenous equivalent, at least as far as these census-based estimates are concerned. While reasons for this are not clear, it may be that the relatively high non-Indigenous figures reflect individuals who are temporarily between jobs. Overall, average Indigenous income is less than half (42%) of the average non-Indigenous equivalent.

Table 4.2. Indigenous and non-Indigenous annual average personal income ($) by labour force status in the Pilbara SD, 2001

	CDEP	Mainstream	Unemployed	Not in labour force	Total
Indigenous	11 041	34 966	11 058	11 811	20 163
Non-Indigenous	n/a	55 226	20 642	14 793	47 624

Source: ABS customised tables.

Table 4.3. Ratios of Indigenous to non-Indigenous annual average personal income ($) by labour force status in the Pilbara SD, 2001

CDEP	Mainstream	Unemployed	Not in labour force	Total
100.0	0.63	0.53	0.80	0.42

Source: ABS 2001 Census customised place of enumeration tables.

It is interesting to compare these average incomes for different parts of the Pilbara as notable variations occur. For example, Table 4.4 shows average incomes from mainstream employment as well as average incomes from all sources for Indigenous and non-Indigenous adults in the four Pilbara SLAs. This reveals that average Indigenous total incomes are much lower in the East Pilbara than in any other part of the Pilbara, and nowhere do they exceed more than 50 per cent of their non-Indigenous equivalents. Ironically, the highest non-Indigenous incomes are also found in the East Pilbara. This just serves to exacerbate the gap in economic status in this region where Indigenous incomes are only 28 per cent of the non-Indigenous level. By far the highest Indigenous incomes from mainstream employment are found in the Ashburton SLA. Also worth noting is the fact that the lowest incomes overall are found in Port Hedland and Roebourne SLAs reflecting their greater share of lower paying non-mining jobs.

Table 4.4. Indigenous and non-Indigenous average incomes from mainstream employment and all sources: Pilbara SLAs, 2001

SLA	Mainstream employment		Total	
	Indigenous	Non-Indigenous	Indigenous	Non-Indigenous
East Pilbara	$30 384	$62 041	$15 833	$55 478
Port Hedland	$36 316	$54 076	$22 396	$46 828
Ashburton	$42 655	$60 699	$22 259	$50 555
Roebourne	$33 380	$50 862	$21 353	$43 792
Total Pilbara	$34 966	$55 226	$20 163	$47 624

Source: ABS 2001 Census customised place of enumeration tables.

Regional prices index

Part of the difficulty in assessing the adequacy of income lies in establishing its purchasing power. This is a complex calculation for which the full range of necessary data inputs usually acquired from a household income and expenditure survey are simply not available. However, one variable for which data are readily available concerns the relative cost of purchasing foodstuffs in the Pilbara compared to the state capital. Since 2001, the Pilbara Development Commission has administered the Pilbara Price Surveillance Scheme to compare grocery and fuel prices in the region with those prevailing in Perth. The surveys are conducted in major supermarkets and so reflect urban prices only. From the perspective of Indigenous community stores and homeland settlements the relativities in prices revealed by these surveys will be absolute minimums. Results from this survey for the period 2001–05 are shown in Table 4.5.

Table 4.5. Index of grocery basket costs in Pilbara towns (Perth = 100), 2001–2005

	2001	2002	2003	2004	2005
Dampier	n/a	102.5	110.1	109.6	111.9
Karratha	104.5	107.6	109.3	107.3	107.5
Marble Bar	n/a	145.5	n/a	136.1	134.4
Newman	n/a	110.6	109.3	109.6	114.6
Onslow	121.7	119.6	122.0	122.3	122.0
Pannawonica	110.6	119.5	116.9	117.7	114.2
Parraburdoo	108.7	121.3	117.1	118.1	109.6
Port Hedland	n/a	113.6	115.8	111.8	111.0
Roebourne	n/a	n/a	129.0	128.5	129.5
Tom Price	107.2	113.2	119.7	115.9	120.8
Wickham	n/a	105.1	111.0	109.5	102.6

Source: Pilbara Development Commission.

Notwithstanding methodological problems in comparing prices between places and over time due to variable coverage, as well as due to some averaging for missing items, it is clear that prices across the Pilbara towns surveyed are consistently higher than prices in Perth. Presently, the regional mark-up ranges from nearly 3 per cent to as much as 34 per cent. Certainly, in areas where the Indigenous population is more heavily represented, including in Roebourne, the relative food price index is generally highest (see Interview segment 23, p. 75).

Welfare income

The actual dollar contribution to regional income from employment and non-employment (welfare) sources in 2001 is shown in Table 4.6. According to these calculations, the total gross annual personal income accruing to the population enumerated in the Pilbara amounted to $1.1 billion. However, only 5 per cent of this ($58 million) went to Indigenous people despite the fact that they accounted for 15 per cent of the adult population. Of greater interest is the fact that only 3 per cent of the total regional income generated by mainstream employment accrued to Indigenous people. The implications of this are reflected in relative levels of non-employment income that, in this context, may be taken as a crude measure of welfare dependency (seeInterview segment 24, p. 75). As much as 28 per cent of total Indigenous income is attributable to non-employment (welfare) sources compared to only 6 per cent of non-Indigenous income. If CDEP income is also counted as welfare income, owing to its notional link to Newstart Allowance, then the level of Indigenous reliance on income from welfare rises to 36 per cent (see Interview segment 22, p. 75).

Table 4.6. Gross annual personal income ($) for Indigenous and non-Indigenous adult residents of the Pilbara SD, 2001

	Indigenous	Non-Indigenous	Total	Indigenous % share of income
CDEP	4 759 040	n/a	4 759 040	100.0
Mainstream	37 239 800	1 033 787 560	1 071 027 360	3.5
Unemployment	2 289 040	9 702 160	11 991 200	19.1
Not in the labour force	13 984 880	58 404 840	72 389 720	19.3
Total	58 272 760	1 101 894 560	1 160 167 320	5.0
Welfare share (exc CDEP)	0.28	0.06		
Welfare share (inc CDEP)	0.36	0.06		

Source: ABS 2001 Census customised place of enumeration tables.

The distribution of income by notional source is also of interest when compared to that recorded for Indigenous people across remote Australia by the 2002 National Aboriginal and Torres Strait Islander Social Survey (NATSISS). This revealed that CDEP accounted for 28 per cent of Indigenous incomes, other wages and salaries accounted for 19 per cent, and government pensions and allowances accounted for 45 per cent. Thus, compared to the average for remote Australia, the Pilbara emerges as quite distinct, with a much lower reliance on CDEP and government payments, and much greater reliance on mainstream wages and salaries. However, as we shall see, there are questions surrounding the real level of dependence on government payments, while the lack of access to administrative data on CDEP wages also raises doubts about that level as well.

One issue that arises in the Pilbara in terms of regional development is the extent to which potential investment dollars earned in the region are lost due to FIFO and other transitory workers remitting their earnings elsewhere (see Interview segment 21, p. 74). The only direct way to estimate this would be to establish the actual incomes of such workers from company records. One indirect way is to recalculate the figures in Table 4.6 using usual residence instead of place of enumeration data. The difference between the two represents a proxy for the income of Pilbara-based non-residents. Overall, this reveals that the usual resident population of the Pilbara earned $169 million less in 2001 than those enumerated there. In terms of income from employment, the shortfall was $127 million (about 12% of regional income from employment). This then provides a crude estimate of lost regional income. Of course, this is likely to be a minimum estimate as an unknown number of usual residents of the Pilbara no doubt also repatriate a component of their earnings. All of this assumes, of course, that individuals enumerated in the Pilbara who were classified as employed actually had their place of employment within the Pilbara. While this is probable, it is not certain.

Centrelink

Because the census-based calculation of non-employment income provides only a proxy measure of dependency on welfare spending, attempts have been made in previous studies to establish a more precise measure of this using administrative data on the composition of welfare income obtained from Centrelink. These data typically produce a higher estimate of welfare income than that generated from census data (Taylor 2004a, 2004b; Taylor & Westbury 2000), and much more in line with the 2002 NATSISS finding that 45 per cent of Indigenous people in the remote Australia are reliant on government payments as their main source of income.

With a view to applying this same method, the amounts paid in benefits (excluding CDEP) to Indigenous Centrelink customers in the Pilbara for a single fortnight in 2005 were requested. Approval for the release of such data are now required by unanimous agreement from three Commonwealth departments—the Department of Employment and Workplace Relations, the Department of Family and Community Services, and the Department of Education, Science and Training—as well as Centrelink. However, on advice from Centrelink, such approval is only presently provided for analyses that are focused on locations or regions that host a COAG Indigenous Communities Coordination Project (ICCP) Trial. As this is not the case for any part of the Pilbara, this important piece of regional economic information is not available.

What are available, however, are the number of Indigenous and non-Indigenous customers classified by type of Centrelink payment received for the fortnight ending March 11, 2005. For reference purposes, these data are shown in full for each SLA within the Pilbara in Table 4.7. It should be noted that customers can receive more than one payment and so aggregate totals shown are of payments, not discrete customers. It should also be noted that Indigenous numbers in these data are as reported to Centrelink by self-identification and the accuracy of such reporting is not known. To render the task of interpretation easier, the data shown in Table 4.7 is distilled to the Pilbara-wide scale in Table 4.8 in order to examine the regional Indigenous share of payments by type.

Table 4.7. Indigenous and non-Indigenous Centrelink customers by payment type, Pilbara SLA, 2005[a]

	Ashburton		East Pilbara		Port Hedland		Roebourne	
	Ind.	Non-Ind	Ind.	Non-Ind	Ind.	Non-Ind	Ind.	Non-Ind
TES[b]								
Austudy	0	*[c]	0	0	0	0	0	0
Newstart	145	134	99	28	252	86	174	96
Newstart mature age	0	*	0	0	*	*	0	*
Partner allowance	*	*	0	*	0	*	8	*
Sickness	0	*	0	0	0	0	0	*
Widow allowance	*	*	*	*	*	*	8	*
Youth allowance	44	38	*	*	56	23	29	29
Total TES	321	224	192	49	396	122	264	147
Pensions								
Age pension	67	395	44	52	79	160	24	124
Carer pension	*	26	*	*	*	*	*	*
Disability support	124	191	67	30	187	126	132	97
Parenting payment single	198	115	82	34	303	107	159	158
Wife pension (Age)	*	*	*	*	*	*	0	*
Wife Pension (DSP)	*	*	*	*	*	*	0	*
Total Pensions	424	746	213	119	620	409	324	386
Other								
Abstudy (>16)	43	*	*	0	71	*	42	*
Abstudy (<16)	84	*	*	0	109	*	82	*
Isolated children	*	128	*	*	*	*	0	26
Child disability	35	80	*	23	40	65	0	0
Orphan pension	*	*	0	0	0	0	*	*
Family tax benefit	361	807	213	245	513	600	326	840
Parenting payment partnered	43	57	48	*	59	29	67	48
PES (Abstudy)	*	*	*	0	*	0	*	*
PES (FaCS)	*	*	0	0	0	*	0	0
Total Other	586	1086	323	286	819	709	546	978
Grand total	1331	2056	728	454	1835	1240	1134	1684

[a]As at March 2005.

[b]TES = Total Employment Services.

[c]* denotes value less than 20.

Source: Centrelink Programming Services, Brisbane.

Thus, in the Pilbara as a whole, Indigenous people account for almost half of all Centrelink payments despite accounting for only 15 per cent of the adult population. More importantly, Indigenous people account for fully two-thirds of all Newstart payments (totalling 670 Indigenous clients), almost 60 per cent of all Youth Allowance payments (129 Indigenous youth), and more than half of all disability payments (510 Indigenous adults). Unfortunately, these data were not made available by sex for confidentiality reasons. However, if we take those on disability pensions alone, we could immediately subtract these from the ranks of those not in the labour force shown in Figs. 3.7 and 3.8 on the assumption that they are not available for work.[1] This represents 30 per cent of the 1723 Indigenous male and females who are not in the labour force, and at a stroke this would reduce the number in this cohort who might theoretically be drawn into the labour force to just 1213.

Table 4.8. Indigenous and non-Indigenous Centrelink customers by payment type: Pilbara SD, 2005[a]

Payment type	Indigenous	Non-Indigenous	Total	Indigenous % of total
Newstart	670	344	1014	66.1
Partner allowance	8	0	8	100.0
Widow allowance	8	0	8	100.0
Youth allowance	129	90	219	58.9
Total NSS	1173	542	1715	68.4
Age pension	314	651	965	32.5
Carer pension	0	26	26	0.0
Disability support	510	444	954	53.5
Parenting payment single	742	414	1156	64.2
Total Pensions	1581	1487	3068	51.5
Abstudy (>16)	156	0	156	100.0
Abstudy (<16)	275	0	275	100.0
Isolated children	0	154	154	0.0
Child disability	75	168	243	30.9
Family tax benefit	1413	2492	3905	36.2
Parenting payment partnered	217	134	351	61.8
Total Other	2274	3059	5333	42.6
Total payments	5028	5261	10 289	48.9
Total Income support payments[b]	3615	2769	6384	56.6

[a]As at March 2005.

[b]Total payments less FBT.

Source: Centrelink Programming Services, Brisbane.

In considering these data, it is important to distinguish Family Tax Benefit (FTB) payments from those that are more clearly designed for income support purposes. The FTB is a complex payment and there are a number of issues that need to be taken into account when using FTB data. For example, FTB customers who receive less than the maximum rate usually have family incomes in excess of $32 500 a year and therefore are not generally considered to be part of the 'welfare' population. However, customers who receive less than the maximum rate because of the operation of the maintenance income test, can have low family incomes and would be considered part of the 'welfare population'. Unfortunately, disaggregation of the Pilbara population in this way was not possible. However, on Centrelink's advice, approximately one-third of FTB customers are also on income support payments and are therefore counted twice in Table 4.8. If we focus solely on income support payments, net of FTB, we can see from Table 4.8 that the Indigenous share of such payments rises to 56.6 per cent. However, this

This is an assumption only and note is taken of the new federal initiatives to encourage workforce participation by those on disability pensions.

share varies across the Pilbara and the data in Table 4.7 indicate that Indigenous customers account for 44 per cent of all income support payments in Ashburton, 71 per cent in East Pilbara, 67 per cent in Port Hedland, and 55 per cent in Roebourne SLAs.

As noted earlier, actual disbursements associated with each of these payments were not forthcoming from Centrelink. Consequently, an attempt is made here to derive a crude estimate of the global amount paid out by Centrelink in the Pilbara by using indicative data on average spending per customer from previous studies of similar remote regions. Thus, in both the East Kimberley and in the Thamarrurr region of the Northern Territory, both of which are not dissimilar to the Pilbara in terms of their profile of Indigenous disadvantage, average spending by Centrelink per customer per fortnight amounted to $300 (Taylor 2004a, 2004c). If we apply this same amount to the customer numbers for the Pilbara we derive a fortnightly Centrelink expenditure of $1.5 million for Indigenous customers, and $1.6 million for non-Indigenous customers, or $3.1 million combined. We can then annualise these amounts simply by multiplying the fortnightly payments by 26. This produces figures of $39 million for Indigenous customers, $42 million for non-Indigenous customers, and $81 million in total. While this latter figure is very similar to the gross figure of $84 million due to persons not employed in Table 4.6, the Indigenous and non-Indigenous amounts are very different with the former being much lower and the latter much higher than the census-based estimates. Ultimately, the fact remains that the actual size of the welfare economy in the Pilbara region remains off the public record. All we can say, based on estimates from public access information, is that it is likely to be in the region of $80 million per annum, with Indigenous people and their families in receipt of at least half of this.

Indigenous perspectives

Interview segment 17

Well young people, like some young married couples, they might need their own space, but the thing is they cannot survive on their own. They are not independent; they still rely on their mum and dad or their grandparents you know. Lot of people are the same. Like that dole money you know, there now, they say, 'why should I go and work?' They stay on that and might pay $50 rent but that's nothing, they comfortable because they got everything. They can live on that. I am on that too, and I am surviving, because I budget my money. I can manage on that, and its not even $400 a fortnight, and I'm on CDEP, plus I got three kids going to school and you know lunch money, plus the running costs for the motor car, 40 kilometres to go to town shopping, fuel, and we have our bills.

Interview segment 18

Well myself, I've always got enough left just before the next pay, you know like families when you lend them the car, it's hard, hard life but you can't say no. So this house is the John Howard house! Everybody comes here when they need or want something, so we are giving giving giving but not receiving. People don't pay us back, which is a problem, we got nothing today, otherwise we gotta wait for pay day. If people ask for the car we give it as long as they put fuel in that's good enough, if they want to go into town, just the fuel. It's a car it doesn't run on air. The other thing for me to get paid (on CDEP) I gotta go and do the hours, 36 a fortnight. I get paid peanuts really.

Interview segment 19

Government has taken over the profile of CDEP and painted it up to be a good thing, but really they are just underpaid, doing good work, but underpaid. I know of someone who moved out of a rental house in January this year, and they charged them $50 an hour for professional cleaners to come in and clean it after. CDEP workers are lucky to get $8 per hour and yet they often do the same sort of work – cleaning and the like. That's cheap labour!

Interview segment 20

Old people have grown up and had to earn their money. They had to feed themselves and they know that. But when all the young people in that house, they all grown up and the nanas and stuff been feeding them so they're dependent on that. Even when they get their own money, when they do the CDEP, because of that dependency they go and blow their own money on whatever they want, but they don't buy the food. They are dependent, they haven't learnt, and when nothing's in the cupboard then nana has to go house to house to get a bit of tucker. And then that nana and them start getting cut short because they can't spin their money out till the next pension, because they starting to feed them. But they are grown up men and women!

Interview segment 21

Another problem for our people is they don't like the idea of fly-in-fly-out taking away jobs from locals. It's not even an Aboriginal non-Aboriginal issue, whitefellas who live here permanent, they like the country, like the place and they want to be permanent but they can't. It's cheaper for the industry and they are building houses for fly-in-fly-out mob. But for people local its bad for business as they take their money away, every ten days they get a week off, they aren't going to spend money here. Lot of local businesses can't cope. When I first came to Tom Price we had three Chinese shops – all gone now there's only one left.

Interview segment 22

You find this anywhere, white or black. Easy street is what our younger generation prefer these days, everything easy, sit on the dole, why work, why go and slave yourself around you see? But sometimes we make em work for it though by going on those survey and heritage for mining with the anthropologists, all working, and that is a hard job! But also it's an education, and a learning process, there's so many hats in that job. Whereas if you put em here just cleaning the maid rooms, which I've been doing, I learn nothing from it, I'm just making an income just to live. But there's nothing from it, I don't benefit from it. You need to get in a good job to get confidence and identity, I reckon, and you need your home base to be good too. So you can enjoy going home, your morale stuff at home, all is good in the home, gets you up in the morning, gets you to work, enjoy your work, enjoy your colleagues, enjoy being a part of a team. And that's not working much around round here!

Interview segment 23

It's an expensive place to live. By the time I buy food, $200–$250 from the store for the week. And then it's gone until the next pay day. I have to go shooting for my meat, can't afford to buy 'im. At the moment I got my family out at the community. I go weekends but they sometimes come and see me. I have to keep the work happening out there, because I can't get anyone to work out there, because a lot of them don't know how to. They'd rather be back here in town putting the music on and you know. Their parents never learn them, and gave them too much of a long rope with their kids and give them money money money. Even CDEP starting to fall down too. If people don't work then we don't pay 'em. But they been seeing others elsewhere who haven't been doing the work and still getting paid, and they are trying to do it here, and they keep trying to do it here. What's happening is that they got no money and they starting to hit the older people, and everybody in the community they start to bludge off.

Interview segment 24

And that's where the problem is, it's not having a nice home, not able to have money that you can access at any day of the week, you have to live off money week by week, still welfare dependent, that we need to get off because that is what this mining is all about. Unfortunately our infrastructures couldn't work that way. Not blaming anybody, but it didn't work and now we gotta look at a way it can work.

5. Education and training

There are two broad perspectives against which the purpose and performance of education in the region may be assessed. The first is culturally-grounded and considers what Indigenous people want from education. According to one analyst, many Indigenous people selectively procure aspects of Western education and ignore others that do not suit their needs and aspirations (Schwab 1998). Consequently, what is desired from education in general, and from schools in particular, can be very different to what these western institutions expect. These desires have been conceptualised in terms of the acquisition of core competencies to deal with the non-Indigenous world, the capacity for cultural maintenance, and access to material and social resources (Schwab 1998: 15). Such acquisitions might be construed as part of an Indigenous social capital model that sees schooling as a means to reinforce capacities within the Indigenous domain.

The second perspective derives from an economic development model and stresses a need to acquire human capital skills in order to participate in the mainstream economy. From this perspective, educational outcomes are measured in terms of participation rates, grade progression, competency in numeracy and literacy skills, and (for the Vocational Education and Training (VET) sector) course completion. Given the focus here on developing a statistical profile of the regional population, consideration is given solely to this second perspective. In part, this is recognition that the human capital model of educational outcomes reflects more closely the platform required if local Indigenous aspirations for enhanced participation in mainstream employment are to be realised. It is also recognition that culturally-grounded educational outcomes are more difficult to quantify and lack readily accessible data sources.

Participation in schooling

A total of 36 schools are located in the Pilbara region administered by three different education sectors – government, catholic, and Aboriginal independent. Collectively, they incorporate five high schools, two colleges, 21 primary schools, five Aboriginal community schools, and two education support centres, along with the Port Hedland School of the Air (Table 5.1). In addition, significant education initiatives built into certain mining agreements, and into the community relations practices of major companies such as Pilbara Iron, Woodside, Dampier Salt, and BHP Billiton, include the provision of scholarships and local educational enrichment programs. Of particular note are the Partnership for Success programs managed by the Graham (Polly) Farmer Foundation at Karratha and Roebourne (Gumala Mirnuwarni) and Port Hedland (Kurtakalku Maya) with outreach at Newman and Tom Price. Table 5.1 indicates the number of Indigenous enrolments in each government and non-government school. In the case of government schools, the number of Indigenous enrolments is shown, along with the Indigenous percentage of all enrolments in parentheses.

Table 5.1. Distribution of government and non-government schools by Pilbara SLAs, 2005

Ashburton SLA	East Pilbara SLA	Port Hedland SLA	Roebourne SLA
Tom Price senior high 33[a] (17)[b]	Rawa independent community school	St Cecilia's college	St Paul's primary
Onslow primary 79 (67)	Parnngurr independent community school	Hedland senior high school 206 (31)	St Luke's college
Paraburdoo primary 15 (6)	Newman senior high school 37 (14)	Port Hedland primary 38 (8)	Karratha senior high 111 (16)
Tom Price primary 52 (15)	Jigalong remote community school 83 (100)	South Hedland primary 248 (77)	Dampier primary 5 (3)
Pannawonica primary 6 (7)	Marble Bar primary 57 (85)	Baler primary 147 (28)	Roebourne primary 253 (98)
North Tom Price primary 34 (13)	Nullagine primary 48 (86)	Cassia primary 87 (25)	Karratha primary 63 (14)
	Newman primary 40 (15)	Port Hedland school of the air 7 (22)	Wickham primary 71 (21)
	South Newman primary 83 (22)	Cassia education support centre	Pegs Creek primary 47 (22)
	Yandeyarra remote community school 58 (100)	Strelley independent community school	Millars Well primary 25 (7)
	Kiwirrkurra remote community school		Tambrey primary 63 (16)
			Karratha education support centre 18 (72)

[a]Number of Indigenous enrolments at the school.

[b]Indigenous percentage of total school enrolments in parentheses.

Source: Western Australia Department of Education.

As can be seen, given the demographic make-up of the Pilbara, many of the government schools are also predominantly Indigenous in terms of their enrolments. Leaving aside the remote community schools, Table 5.1 reveals other major concentrations of Indigenous students throughout the Pilbara school system. It also shows some notable voids. For example, Roebourne primary is essentially an Indigenous school, with Indigenous children accounting for 98 per cent of its enrolments. The primary schools at South Hedland, Marble Bar, Nullagine and Onslow also have high percentages of Indigenous children enrolled. Schools where Indigenous students are more notable for their relative absence include Paraburdoo, Pannawonica, North Tom Price, Millars Well, Tambrey and Dampier. At the high school level, as much as one-third of enrolments at Hedland senior high are Indigenous, providing it with a significant Indigenous profile, certainly compared to Karratha or Tom Price.

In terms of the two perspectives on educational purposes and outcomes posited above, the existence of an Aboriginal independent community sector in the Pilbara is also significant and flags an important socio-economic distinction between populations in the (mostly) coastal towns and interior Aboriginal communities. All of these independent schools (Strelley has annexes at Woodstock and Warralong), are bilingual, with Aboriginal teachers trained on site to teach in the vernacular. They are run by school committees comprised of representatives from local family groups. Their aims (and outcomes) are cast much more in terms of the Indigenous social capital model posited above. For example, the aims of Strelley school include the teaching of survival skills, community involvement, maintenance of Nyangumarta traditions, learning about Aboriginal and non-Aboriginal cultures, self-identity, and keeping the school and children close to parents. Inevitably, this produces a different set of outcomes to those sought from the mainstream education system with its greater stress on English literacy, numeracy, individual achievement and mobility. Thus, the distribution of student enrolments by school type provides some measure of the relative importance of social versus human capital outputs from the regional education system.

Composition of enrolments

In the second semester of 2004, a total of 6374 students were enrolled in Pilbara schools between Years 1 and 10, which approximates the compulsory school age range of 5–17 years. Of this number, 1684 were Indigenous students representing 26 per cent of the total compulsory enrolment. An additional 672 students were enrolled in Years 11 and 12, and 127 (19%) of these were Indigenous. While Indigenous population estimates by single year of age do not exist for 2004, if the Indigenous ERP for broad age groups provides any guide these enrolment levels would yield high rates at close to 100 per cent, at least for the compulsory school years.

Most Indigenous (and non-Indigenous) enrolments are at schools located in the Port Hedland and Roebourne SLAs as indicated in Fig. 5.1. Also shown is the fact that enrolments start to fall away quite sharply in all regions as the secondary school years progress, especially beyond age 15. As a consequence, in the second semester of 2004, only 21 Indigenous males were enrolled in Year 12 (with none in Ashburton schools), and only 24 Indigenous females. Indigenous students comprised just 9 per cent of all Year 12 school enrolments.

Figure 5.1. Indigenous school enrolments in Pilbara SLAs by single year of age, 2004

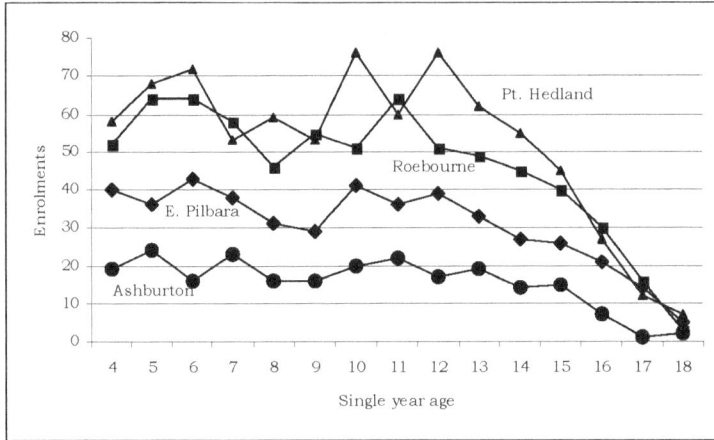

Source: Western Australia Department of Education and Training

Figure 5.2. Ratio of male to female Indigenous school enrolments in Pilbara schools by single year of age, 2004

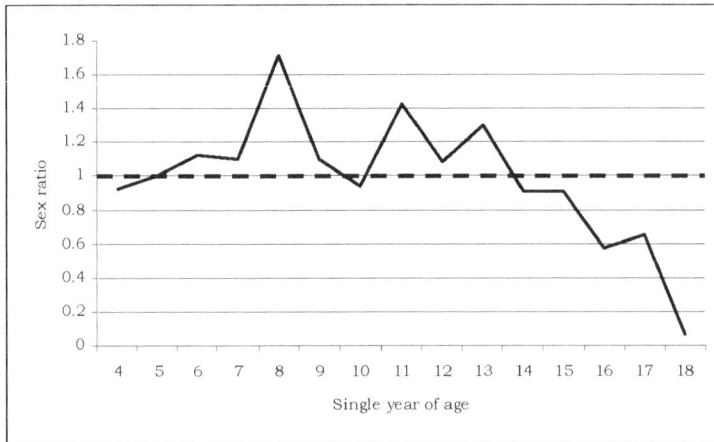

Source: Western Australia Department of Education and Training

One reason for this reduced Indigenous participation in formal education appears to be a relatively higher attrition among male students. Fig. 5.2 shows the sex ratio of Indigenous enrolments by single-year age group. The higher ratio of males in the 5–14 age groups is broadly consistent with the ERP sex ratio for those ages (1.08 for the 5–9 age group and 1.09 for the 10–14 age group), but at older school ages the ERP sex ratio is much closer to parity (0.99) than the observed school enrolment sex ratio. This suggests that Indigenous males drop out of schooling at a higher rate than their female counterparts after age 14.

Retention rates

Table 5.2 shows apparent retention rates for Indigenous students in Pilbara schools from Year 8 to Year 10, and from Year 10 to Year 12. These rates are compared with those recorded for Indigenous students generally in Western Australia, as well as with all non-Indigenous students in the state. The state-wide data are shown for 2003, this being the latest year for which data are available. The rates represent the proportion of those previously in Year 8 who were retained by Year 10 two years later (2001 to 2003 for the state-wide figures, and 2002 to 2004 for the Pilbara figures). The same calculation is made in respect of those previously in Year 10 who were retained by Year 12.

Table 5.2. Apparent retention rates for Indigenous and non-Indigenous students in Western Australian and Pilbara schools: 2003 and 2004

	Apparent retention rates	
	Year 8 to Year 10	Year 10 to Year 12
WA Indigenous (2003)	93.2	29.3
WA Non-Indigenous (2003)	98.6	72.5
Ashburton Indigenous (2004)	65.0	22.2
Ashburton Non-Indigenous (2004)	71.7	48.7
East Pilbara Indigenous (2004)	105.0	48.4
East Pilbara Non-Indigenous (2004)	87.5	53.7
Port Hedland Indigenous (2004)	80.7	39.5
Port Hedland Non-Indigenous (2004)	75.2	56.4
Roebourne Indigenous (2004)	102.4	35.5
Roebourne Non-Indigenous (2004)	85.3	60.4

Sources: Western Australia Department of Education and Training; Steering Committee for the Review of Government Service Provision (SCRGSP) 2005: Table 3A:22, 32.

Taking the state-wide situation first, this reveals that for all students in Western Australia, both Indigenous and non-Indigenous, retention rates from Year 8 to Year 10 are high and generally close to 100 per cent. In the Pilbara regions, however, the retention of Indigenous students is more varied with only 65 per cent retained in Ashburton and 81per cent in Port Hedland, while the rates in East Pilbara and Roebourne suggest problems with numerator/denominator concordance. As a general observation, retention to Year 10 is lower in the Pilbara than in Western Australia as a whole. When it comes to school retention beyond the compulsory years from Year 10 to Year 12, Indigenous youth generally in Western Australia tend to opt out, and the Pilbara is no exception. Once again, schools in the Ashburton SLA record very low Indigenous retention rates. From a labour market perspective, retention to Year 10 is a significant step for Indigenous students with evidence from the 1994 NATSIS indicating that it almost doubles the chances of Indigenous employment. Further retention to Year 12 increases these chances further still, though more so for females (ABS & CAEPR 1996: 70–5).

The impact of these retention rates is reflected in census data on the highest levels of schooling completed as reported by all adults (those over 15 years). These levels are shown in Table 5.3 for Indigenous adults in the Pilbara, while figures for non-Indigenous adults are also provided for comparative purposes. As much as 10 per cent of the Indigenous adult population of the Pilbara has not been to school. From a labour market perspective, this may not be significant as these people are concentrated in older age groups and reflect the legacy of past exclusionist policies.

Table 5.3. Highest level of schooling completed: Indigenous and non-Indigenous adults[a] in the Pilbara SD, 2001

	Indigenous			Non-Indigenous		
	Males	Females	Total	Males	Females	Total
Year 8 or below	20.2	15.8	18.0	5.6	5.0	5.4
Year 9 or equivalent	13.0	11.8	12.4	7.0	6.0	6.5
Year 10 or equivalent	31.8	35.0	33.4	35.9	31.9	34.2
Year 11 or equivalent	12.3	12.8	12.6	16.1	15.6	15.9
Year 12 or equivalent	13.2	12.3	12.7	35.0	41.1	37.6
Did not go to school	9.5	12.2	10.8	0.5	0.4	0.4
Total	100.0	100.0	100.0	100.0	100.0	100.0

[a]Refers to population 15 years of age and excludes highest level of schooling not stated.
Source: ABS 2002b.

Of more interest are the relatively low proportions who have completed Year 10 or above. As much as 88 per cent of non-Indigenous adults counted in the Pilbara finished school at Year 10 or above, with 54 per cent completing beyond Year 10. The comparable figures for Indigenous adults are just 59 and 25 per cent. In terms of current numbers (2005), and assuming that these census rates still hold, this would mean that an estimated 571 Indigenous adults in the Pilbara would have completed schooling to Year 12. No doubt most of these are already currently employed, meaning that most of those who might be sought for employment positions would have school attainment levels below Year 12. As numerous studies have shown (Daly 1995; Hunter 1996, 2004; Hunter & Schwab 2003), this contrast in levels of schooling completed is highly significant in terms of explaining overall differentials in the rate of Indigenous and non-Indigenous participation in mainstream employment.

One very tangible response to this situation has been the establishment of the Gumala Mirnuwarni and Kurtakalku Maya Partnership for Success programs in conjunction with select students and their families in Karratha, Roebourne, Port Hedland, Newman and Tom Price. In addition to these, Woodside supports the Warrgamugardi Yirdiyabura 'pathways to employment' program at Roebourne. The aim of these programs is to prepare students to compete for employment, apprenticeships, traineeships and/or tertiary entrance after leaving school. To achieve this, students with identified capacity to complete school are targeted and then provided with a range of intensive support structures including trained mentors, access to after school hours support and a school resource centre/library for study, a comprehensive leadership/study skill program from Year 8 to Year 12, a full-time program coordinator, family and home support, industry support, and access to a tertiary motivational program.

The Gumala Mirnuwarni program at Karratha and Roebourne was the first of these initiatives, commencing in 1997. By 2004, a total of 110 students had participated with 31 of these still in the program. The Port Hedland-based Kurtakalku Maya program started in 2002 and has had a similar up-take rate, with 46 students involved up to 2004 and 21 enrolled as at 2004. While some evidence exists of increased high school retention rates as a consequence of these programs (Graham (Polly) Farmer Foundation 2004), the telling information concerns post-school outcomes. Of the 104 students who have been through these programs, 15 are now enrolled in University or Technical and Further Education (TAFE) programs, 21 are in trade apprenticeships, and 10 have found direct employment (McCorry 2004). Presumably, the impact of these initiatives is incorporated in the aggregate profiles presented here, and while they clearly demonstrate positive outcomes for those students involved it has to be said that, numerically, their overall impact to date is limited. However, to the extent that these initiatives continue to raise Indigenous educational participation and outcomes and, more importantly, to the extent that they expand in terms of student numbers participating, then they are likely with time to contribute to an overall positive shift in aggregate indicators.

With regard to such initiatives, there is clearly a sense among some local Indigenous people of a need for some targeting of educational resources, particularly towards younger age groups, to ensure that future school leavers are equipped with a skill-set that will widen their life choices (see Interview segment 12, p. 61; Interview segment 27, p. 93; Interview segment 29, p. 93). Others see such targeting as too narrowly focussed (see Interview segment 10, p. 60).

Attendance

For most schools in the Pilbara, the educational impact of relatively low levels of Indigenous school enrolment is compounded by low Indigenous school attendance (Table 5.4). According to these data, in all parts of the Pilbara, less than 80 per cent of Indigenous children enrolled in primary school years actually attend school on a regular basis, with this figure falling to as low as 60 per cent in East Pilbara schools. In Years 8 to 10, the rates are even lower with fewer than half of enrolled students in Ashburton attending. These rates are far lower than those recorded for non-Indigenous students in Pilbara schools.

Table 5.4. Indigenous and non-Indigenous attendance rates in Pilbara schools: primary and secondary years, 2004[a]

	Indigenous Yr 1–Yr 7	Non-Indigenous Yr 1–Yr 7	Indigenous Yr 8–Yr 10	Non-Indigenous Yr 8–Yr 10
Ashburton	65.4	87.9	43.3	85.9
East Pilbara	59.9	91.5	54.1	89.0
Port Hedland	77.3	92.8	75.1	90.3
Roebourne	72.1	92.5	60.6	92.0

[a]Second semester, per cent of enrolments attending on a regular basis.
Source: Western Australia Department of Education.

We can use the attendance rates shown in Table 5.4 against the numbers enrolled in each school year to produce estimates of numbers actually attending classes (Table 5.5). This is an important device when set against the labour demand and supply issues outlined in chapter 3. Thus, in terms of potential Indigenous labour supply emanating from the Pilbara school system in the next couple of years, only 286 individuals have been in regular school attendance, with a large proportion of these located in schools in the Port Hedland Shire.

Table 5.5. Estimates of Indigenous and non-Indigenous students attending Pilbara schools: Primary and secondary years, 2004[a]

	Indigenous Yr 1–Yr 7	Non-Indigenous Yr 1–Yr 7	Indigenous Yr 8–Yr 10	Non-Indigenous Yr 8–Yr 10
Ashburton	83	553	20	106
East Pilbara	161	375	50	136
Port Hedland	363	905	126	290
Roebourne	263	1354	90	566
Total	870	3187	286	1098

[a]Second semester.
Source: Western Australia Department of Education and Training.

All of these official data and estimates regarding school access and participation are based on averages. What they do not show, and what would be more important to reveal (although it is well-nigh impossible), are the day-to-day levels of individual attendance at school. Given the variability in attendance and high levels of short-term population mobility among the Indigenous population it cannot be assumed that aggregate data refer consistently to the same individuals. Since children often accompany adults in their movements across, into and out of the region it seems likely that some mobile children may be overlooked as part of the regular school population. Moreover, since attendance registers are taken each morning, no records exist regarding student participation beyond morning sessions. The prospect thus exists that the attendance rates presented here, especially those for Indigenous students, are overly-favourable. Certainly, some locals consider school participation and attendance levels to be less than satisfactory (see Interview segment 25, p. 00; Interview segment 28, p. 00).

Outcomes

As already noted, from the standpoint of participation in regional economic development, educational achievement is a key prerequisite. While studies reveal a clear positive relationship between economic status and level of educational achievement (as measured by standard indicators such as highest level of schooling completed, and post-school qualifications), an important shortcoming is their lack of measurement of the quality of education outcomes. For example, age at leaving school or highest level of schooling completed does not necessarily equate with school-leaving grade level achievement. In fact, for many Indigenous students in remote areas, age or grade level is a poor indicator of achievement as many Indigenous students perform substantially below their age and grade levels in terms of literacy and numeracy competencies. Thus, while data on participation in the education system provide an important indication of access and utilisation, it should be noted that they are less revealing about outcomes in terms of demonstrated ability, no matter from what perspective this might be measured.

In Western Australia, outcomes from education are measured using benchmarks devised by the Western Australian Literacy and Numeracy Assessment program (WALNA). This is a curriculum-based assessment that tests students' knowledge and skills in numeracy, reading, spelling and writing. The WALNA test is administered annually to all students in Western Australian schools (including Catholic schools) in Years 3, 5 and 7, although a few exemptions are made. The test gathers information on the performance of school children in relation to nationally agreed benchmarks in numeracy, reading, spelling and writing, and in relation to that of other Year 3, 5 or 7 students across Western Australia. The national benchmark standard is an agreed standard of performance that professional educators across the country deem to be the minimum level required for students at particular key stages in their educational development in order to make adequate progress. By providing an indication of how students are faring against the national benchmark and in relation to state performance, the WALNA assessment assists in identifying those students who would benefit from extension, as well as those not meeting the minimum expected standard.

Unfortunately, the Western Australia Department of Education advises that the relatively small numbers of Indigenous students who sit for these tests in Pilbara schools prevents the construction of reliable estimates of Indigenous student achievement specifically for that region. As a consequence, it is not possible to establish precisely the number of Indigenous students within the Pilbara school system who are likely to progress with, or without, difficulty towards an outcome that would satisfy the requirements for a successful engagement with the mainstream labour market (at least as determined by benchmark achievements). However, for the first time, the 2004 National Report on Schooling reports estimates of the proportion of students achieving benchmark scores at the Ministerial Council on Education, Employment, Training and Youth Affairs (MCEETYA) geolocation level, and the data for reading achievement for Western Australia are shown in Table 5.6. In this classification, the Pilbara falls entirely within the 'very remote' category, except for Karratha and Port Hedland which fall under the 'remote' category.

Table 5.6. Proportion of WA students achieving the national benchmarks in reading: Total and Indigenous populations and MCEETYA geolocations, 2004

	Year 3	Year 5	Year 7
All students[a]	95.6% ± 1.4%	93.7% ± 1.0%	88.9% ± 1.1%
Indigenous students	84.1% ± 5.0%	74.2% ± 3.9%	57.6% ± 3.9%
Metropolitan students	96.4% ± 1.2%	94.9% ± 0.9%	91.0% ± 1.0%
Provincial students	95.1% ± 1.9%	92.4% ± 1.3%	87.0% ± 1.6%
Remote students	92.0% ± 3.2%	90.1% ± 2.3%	81.4% ± 2.6%
Very Remote students[b]	85.7% ± 4.9%	76.8% ± 4.7%	59.4% ± 4.7%

[a]All Year 3, 5 and 7 students in Western Australia tested in both government and non-government schools in 2004, plus students who are exempt from testing (education support students) who are classified as not meeting the benchmarks

[b]Very remote category includes the Pilbara except for Karratha and Port Hedland which are included in 'remote'.

Source: Western Australia Department of Education and Training.

While the data shown by geolocation are for all students, the Pilbara (very remote) estimates appear to align quite closely with the sState-wide estimates for achievement among Indigenous students. Ultimately, the true levels for Indigenous students in the Pilbara remain unknown, but we can assume they do not exceed those implied in Table 5.6. Thus, for Indigenous students in Pilbara schools outside of Karratha and Port Hedland between 80.8 per cent and 90.6 per cent achieve Year 3 national reading benchmarks. By Year 5 this range had fallen to between 72.1 per cent and 81.5 per cent, and by Year 7, somewhere between just over half and two-thirds of all students (54.7% and 64.1%) were achieving national benchmarks (see Interview segment 30, p. 00). If the rates shown for 'remote' schools apply to the Karratha and Port Hedland student body, then achievement in those places would be higher, but still ranging between 78.8 per cent and 84 per cent by Year 7. Whether these same rates are achieved by Indigenous students in these locations remains unknown, at least in terms of publicly-available information.

Participation in vocational education and training

School-based and post-secondary education and training leading to the acquisition of formal workplace qualifications is available in the Pilbara from a variety of public and private providers, ranging from the Pilbara College of TAFE (with delivery points at South Hedland, Karratha, Roebourne, Wickham, Tom Price,

Newman, Paraburdoo, Pannawonica, Pundulmurra, Roebourne Regional Prison, Cheeditha community, Onslow, Jigalong, Port Hedland, Boodarie and across discrete communities), to industry-based training providers such as ATAL based in Dampier.

Table 5.7 shows the number and proportion of Indigenous and non-Indigenous enrolments in TAFE courses in the Pilbara by course level in 2004. These data refer to publicly funded providers (TAFES and universities) as well as private providers receiving public funds. Enrolment data for private providers undertaking VET activity on a fee-for-service basis are not collected by the Department of Education and Training. Overall, a total of 4740 enrolments were recorded, with Indigenous enrolments accounting for just over one-third (36%) of these. Indigenous males are more represented than Indigenous females in the TAFE sector, with Indigenous males accounting for 40 per cent of all male enrolments, and Indigenous females accounting for 31 per cent of all female enrolments. Indigenous males and females are more likely than their non-Indigenous counterparts to be enrolled in short miscellaneous enabling courses with no formal certification attached (11% and 15%, compared to 6% and 11%). Also evident is the fact that Indigenous enrolments are concentrated in Certificate level I and II courses, while non-Indigenous enrolments are far more likely to be in Certificate levels III and IV. One variation from this pattern is the slightly higher proportion of Indigenous males enrolled in diploma courses, although this only amounts to 38 persons.

Table 5.7. Indigenous and non-Indigenous VET enrolments by course level: Pilbara SD, 2001[a]

Course level (see key)	Indigenous				Non-Indigenous			
	Males		Females		Males		Females	
	No.	%	No.	%	No.	%	No.	%
No level	120	11.2	97	15.6	106	6.5	152	10.8
1	432	40.3	184	29.6	164	10.0	170	12.1
2	333	31.1	205	33.0	608	37.1	437	31.1
3	142	13.2	101	16.2	473	28.8	467	33.2
4	7	0.7	27	4.3	247	15.1	133	9.5
5	38	3.5	8	1.3	43	2.6	46	3.3
Total	1072	100.0	622	100.0	1,641	100.0	1,405	100.0

[a]Excludes Indigenous status not stated. Includes all VET enrolments collected by the Western Australia Department of Training from publicly funded providers (TAFE colleges and universities) and from private providers receiving public funds. Enrolment data for private providers undertaking VET activity on a fee-for-service basis are not collected by the Department of Training.

Key: 1. Certificate I; 2. Certificate II; 3. Certificate III; 4. Certificate IV; 5. Diploma.

Source: Western Australia Department of Education and Training.

For some age groups, the rate of enrolment in VET courses is very high. Table 5.8 shows the enrolment rate by broad age-group in 2001 and reveals that almost three-quarters of Indigenous youth aged 15–19 were enrolled in VET courses, almost two-thirds of those aged 20–24, and half of those aged 25–29. Even in prime working-age groups, up to age 50, participation remains reasonably high. Overall, the Indigenous participation rate in 2001 was around 39 per cent, which is similar to the rate for 2004.

Table 5.8. Indigenous VET enrolments by broad age group: Pilbara, 2001

Age group	Enrolments	Per cent of age group[a]
10–14	76	9.7
15–19	430	71.7
20–24	331	61.6
25–29	303	50.4
30–39	419	40.7
40–49	228	32.0
50–64	80	15.4
Total	1867	39.0

[a]Based on ERP.

Source: Western Australia Department of Education and Training.

Outcomes

To measure performance in the VET sector, the Western Australian Department of Education and Training has identified a number of key performance measures relating to efficiency, effectiveness and quality. In relation to the effectiveness of the training system, the key indicator is the rate of successful completion of modules – the components from which courses are constructed. Table 5.9 compares the rates of successful module completion for Indigenous and non-Indigenous males and females enrolled in Pilbara training courses in 2004. Clearly, outcomes for Indigenous females are the least favourable. Less than two-thirds of enrolled Indigenous females successfully completed their module, with 37 per cent failing or withdrawing before completion. Indigenous males performed somewhat better, with only 29 per cent failing or withdrawing, although compared to non-Indigenous males (who had a 94% success rate) this was a very poor outcome. Nonetheless, in terms of regional labour demand and supply, these data indicate that substantial numbers of Indigenous people in the Pilbara, and especially males, are participating and achieving in vocational education, mostly at Certificate I and II levels.

Table 5.9. Indigenous and non-Indigenous VET module outcomes in the Pilbara, 2004[a]

Outcome[b] (see key)	Indigenous				Non-Indigenous			
	Males		Females		Males		Females	
	No.	%	No.	%	No.	%	No.	%
1	2908	62.0	1576	55.3	5731	80.1	3,718	71.7
2	3	0.1	16	0.6	707	9.9	580	11.2
3	403	8.6	192	6.7	288	4.0	464	9.0
4	655	14.0	482	16.9	116	1.6	299	5.8
5	718	15.3	583	20.5	316	4.4	123	2.4
Total	4687	100.0	2849	100.0	7158	100.0	5,184	100.0

[a]Excludes those enrolled in modules who are continuing studies into the next collection period and Indigenous status not stated.

[b]Categories 1–3 represent successful outcomes.

Key: 1. Competency Achieved/Pass; 2. Recognition of Prior Learning; 3. Non-Assessable Enrolment – Satisfactorily Completed; 4. Competency Not Achieved/Fail; 5. Withdrawn.

Source: Western Australia Department of Education and Training.

The module load completion rate (MLCR) provides another measure of performance, and with this indicator it is possible to compare the Pilbara with data for all of Western Australia (Table 5.10). The MLCR represents the sum of student curriculum hours for successfully completed modules expressed as a proportion of the total student curriculum hours across all module enrolments. In 2004, this rate was only 58 per cent for Indigenous module enrolments in the Pilbara – less than two-thirds of the level reported for non-Indigenous students in the Pilbara, though more favourably placed in regard to the overall Western Australia average.

Table 5.10. Indigenous and non-Indigenous average MLCR (%): Pilbara and Western Australia, 2004

Pilbara Indigenous	Pilbara Non-Indigenous	Western Australia total
58.0	82.9	72.7[a]

[a]2001 figure.

Source: Western Australia Department of Education and Training.

Qualifications

A key human capital requirement in the regional labour market, and a primary product of the education and training system, is the acquisition by individuals of formal qualifications. While program data can reveal numbers passing through courses, it remains the case that the five-yearly census provides the most comprehensive source of data on the number of individuals within the region who are likely to hold post-secondary qualifications.

At the 2001 Census, a total of 11 705 adults in the Pilbara reported having some form of post-school qualification, but Indigenous adults accounted for just 3 per cent of these. With 43 per cent of all adults holding a post-school qualification, the population of the Pilbara is relatively skilled (the equivalent figure in Western Australia as a whole is 39%), although this clearly does not apply to the Indigenous population. Table 5.11 shows the distribution of Indigenous and non-Indigenous non-school qualifications by qualification level. Fully 88 per cent of Indigenous adults hold no qualification compared to 53 per cent of non-Indigenous adults, although males tend to be more qualified than females. Of those with qualifications, Indigenous people are far less likely to have diplomas and degrees.

Table 5.11. Percentage distribution of Indigenous and non-Indigenous adults in the Pilbara with non-school qualifications, 2001[a]

Non-school qualification	Indigenous			Non-Indigenous		
	Males	Females	Total	Males	Females	Total
Postgraduate Degree	0	0	0	1.4	0.7	1.1
Graduate Diploma and Graduate Certificate	0	0	0	0.8	1.9	1.3
Bachelor Degree	0.5	1.6	1.1	8.0	11.4	9.4
Advanced Diploma and Diploma	0.9	2.3	1.6	5.0	7.0	5.9
Certificate	12.6	5.8	9.0	40.8	13.1	29.2
No qualification	86.1	90.3	88.3	44.0	66.0	53.2
Total	100.0	100.0	100.0	100.0	100.0	100.0

[a]Excludes non-school qualification not stated.
Source: ABS 2002b.

If we assume that these 2001 rates remain constant, we can produce an estimate of the numbers in the resident Indigenous adult population by 2006 (4760) who would hold a qualification. This yields a figure of 557 with a qualification, 428 of whom would have certificate level, 76 diploma level, and 52 degree level, although these are minimum estimates as they are based on rates calculated net of individuals who did not indicate their qualification status. Overall, this 2006 level would represent an increase of 60 Indigenous adults since 2001 with a qualification. Set against the scale of output from the VET sector, and the numbers passing through mining company traineeships, this estimated increase in qualified individuals appears far too low and suggests that the overall Indigenous regional rate of post-school qualification is likely to have risen since 2001.

Differences are also evident in the field of qualification reported, both by sex and by Indigenous status (Table 5.12). Although high non-response to the census question on field of qualification undermines the quality of the data, it appears that most qualifications held by Indigenous males are in engineering and building, similar to their non-Indigenous counterparts. Among both Indigenous and non-Indigenous females, on the other hand, qualifications in health, education and management predominate. These differences in field of qualification are broadly in line with occupational variations already highlighted between males and females, regardless of Indigenous status.

Table 5.12. Non-school qualification by field of study: percentage distribution of Indigenous and non-Indigenous males and females in the Pilbara SD, 2001[a]

Non-School Qualification (see key)	Indigenous			Non-Indigenous		
	Males	Females	Total	Males	Females	Total
1	1.4	0.0	0.7	3.2	3.0	3.1
2	0.0	0.0	0.0	0.6	0.9	0.7
3	54.9	1.6	29.9	64.2	5.1	44.9
4	14.0	1.6	8.1	11.1	0.5	7.6
5	6.0	2.1	4.2	2.1	2.1	2.1
6	7.0	20.5	13.3	2.4	20.4	8.3
7	2.3	12.6	7.2	3.2	19.4	8.5
8	1.9	37.4	18.5	5.1	25.3	11.7
9	5.6	12.1	8.6	2.5	9.9	4.9
10	4.2	4.7	4.4	0.9	2.4	1.4
11	1.4	4.2	2.7	3.5	10.5	5.8
12	0.0	1.6	0.7	0.0	0.0	0.0
13	1.4	1.6	1.5	1.0	0.5	0.8
Total	100.0	100.0	100.0	100.0	100.0	100.0

[a]Excludes non-school qualification not stated.
Key: 1.Natural and Physical Sciences; 2. Information Technology; 3. Engineering and Related Technologies; 4. Architecture and Building; 5. Agriculture, Environmental and Related Studies; 6. Health; 7. Education; 8. Management and Commerce; 9. Society and Culture; 10. Creative Arts; 11. Food, Hospitality and Personal Services; 12. Mixed Field Programmes; 13. Field of Study inadequately described.
Source: ABS 2002b.

Given the relative skills profile of Indigenous people portrayed here, a question arises as to whether they will have the requisite qualifications necessary to assume positions within the expanding Pilbara labour market. At one level, this requires knowledge of future labour demand, both in terms of size and composition. One set of estimates that provides this, though with heavy caveats as previously indicated, is available from the December 2004 version of the biannual labour market forecasts produced by the Centre of Policy Studies at Monash University (Table 5.13). These suggest that growth will occur mostly at the top and the bottom of the skills range, which is encouraging for Indigenous employment prospects given the relatively low Indigenous skills profile, but potentially constraining in terms of subsequent advancement into higher occupational levels. At the very least, it points to the need for continuing on-the-job training for Indigenous workers. Equally, there appear to be no illusions about the limitations of training opportunities given the depth of prevailing disadvantage in terms of basic human capital skills and the circumstances in which many Indigenous youth and families find themselves (see Interview segment 10, p. 00; Interview segment 16, p. 00; Interview segment 34, p. 00; Interview segment 57, p. 00).

Table 5.13. Estimates of labour demand by qualification: Pilbara SD 2004/5–2011/12

Qualification	2004/05	20011/12	Change	%
Post-graduate degree	342	450	108	4.4
Graduate diploma	474	603	129	5.2
Bachelor degree	2348	2982	634	25.6
Diploma	1567	1876	309	12.5
Certificate III or IV	4936	5220	284	11.5
Certificate I or II	227	279	52	2.1
Year 12	4932	5331	399	16.1
Year 11 and below	7639	8196	557	22.5
No educ attain	18	19	1	0.0
All qual. levels	22 483	24 956	2473	100.0

Source: Centre of Policy Studies, Monash University.

Indigenous perspectives

Interview segment 25

If I am sick of town I can close this house and get back to the block, but we got four kids at school, for my two daughters living here. We are trying hard for Aborigine kids to go to school. We gotta need that, we gotta put our children on their toes, I tell 'em to 'go to school, go to school, don't be ducking around', some kids might be just ducking around the gully here, and when they come back they say, 'I been to school', but it's not, they bin hiding in the bush here! The school comes to see you if your kids been missing, and they sit down and talk to the kids and really tell them, 'you must go to the school!'

There's no school bus, nothing, they only got school bus going to Wakathuni, but they should do the rounds and do town too. My daughter asked for that, she went to the meeting. They reckoned funding, but you'd think the mining company could donate something to the school you'd think they would be rich enough, eh? Don't know if anybody asked them. Surely they would have a spare bus that they could donate, and they could take the kids on excursions and things like that. But a lot of kids these days haven't got a chance of going to school, because of drugs and alcohol.

Interview segment 26

There is a school bus that comes through. Getting kids to school is one of the few things that works around here. The buses aren't too worried about taking other people from the community to work. Now and again someone will get on the school bus if they miss a lift. And its really good for kids to see people going off to work.

Interview segment 27

I think something has to be attached to not so much everyday school, but something attached to the school and the first step into the Gumala Mirnuwarni program. That's one area, because we can't give much to our 20 to 30 year olds. There is a limit to what you can give to them as they have already gone through the gate. But the 6 to 13 years of age you can start developing, and as they develop they will come back to what is hoped to be a fairly stable family life. The horse has bolted in the other age groups. And the generation gap is too big. You need to develop parenting and work ethics now.

Interview segment 28

Lots of our kids are dropping out of school. A lot of them think, 'well what's the point?' They don't think there is any need for school, and a lot of peer pressure and that sort of thing. Ganja [marijuana] and alcohol are a real problem with young people who should be at school. If you go around South Hedland about 11 pm you'll see heap of young boys and girls out on the street smoking ganja and fighting, young people. The Government has forgotten about them people, its sad real sad.

Interview segment 29

Most of the key to the future lies with young people, because they are the generation who will benefit from increased employment in this region. If they aren't ready, or have no commitment towards changing, its pretty obvious what the future will be like. Some 14 year-olds are already just looking towards their pension. They are limited in what they can feed into. Woodside and Hamersley Iron have done some work with those sort of age groups and they given them skills and employment, but their percentages of employed Aboriginal people are still really small. But it's not impacting on young kids. If we don't get that generation ready there will be a big gap which is already showing.

Interview segment 30

One of the other obstacles for our kids is education. Accessing education and the way in which education is being provided and all the tests that kids have to do to keep with the national level. Most Aboriginal kids can't get up there, they are lower than the benchmark. It's the way it's set up, a communication breakdown, and because there is policy set up by the government and Aboriginal people don't have any input into those things. All the policy is done down there in Perth and no community people are involved. They don't know what it's like out here. And they make a lot of recommendations for education but it's not including the community itself. There should be more local input into education, and for older Aboriginal people to be involved as an advisory council. People in the south have a different perspective and different ideas.

Interview segment 31

The reason kids aren't getting through the education system is social stuff, social issues, basically that's what it is. All that social economic problems, like housing, appropriate education. Well for example – I worked out at a small community for many years, I helped to administer that community. I worked a lot with the education the X society who are full on with their education, but language was first, English was second. Now that I've grown old and I've seen them children grow older, they didn't learn nothing, because they forgot that they have to exist out in the urbanised area, as well as in their community. And that's when the problem came, there were also problems with families who didn't know how to budget their money, didn't know how to maintain their Homeswest home, all those issues. Putting language first was a big problem, language should be born and bred into us, it's not part of a learning process. And those are the traditional people I am talking about. The urbanised ones are all fine because they've had that education opportunity, but traditional people haven't and they are the ones that matters most of all in a mining world, because they're the first ones.

Interview segment 32

Access to education the main barrier. At the community, I am the teacher, but I'm not qualified yet. If I get through the certificate 3 they will fast track me to a degree. It will be four years of study off campus, I can't leave my family. I want my kids to get a good education so they can get a good job. My son has a really good job. He is working with Newcrest mining. He's still training. I got six kids and he is the eldest. He helps me out with some money too.

Interview segment 33

Law and culture is important, same like education. We just got back from law meeting at Jigalong. Its good it still happens, and its going to be around for a long way down the track with all the young fellas coming through. At the moment I am writing down all the songs, and words of our songs.

Interview segment 34

I see there are a lot of opportunities by Pilbara Iron for traineeships, and that's the only way we can get them, but because three-quarters of them are illiterate, that's where the downfall is. Being illiterate and having drug and alcohol problem because of their home base. I reckon it's the home base that creates all of those. If you gave a good home environment to one of those boys or girls, jeez, you know, they could blossom and flourish and be quite intelligent. You have to give some confidence to people to look at what needs to be changed in their community. See, over the years when I have worked with community people, well before any infrastructure was placed on their land, some of them weren't given any decisions to say, 'well this is the type of house that we want and these are the type of things we need'. It was just like Homeswest, or State Housing as it was in those days, just went bang 'here it is', and bang 'there you go'. And there are these people left standing there going, 'well I come from the desert I don't know anything about this'. That's where the social problem, lack of confidence and low self esteem that all came into play. It's like anybody, give them a good home environment and God they feel good about themselves and get up in the morning and say, 'great my home is clean and I've got a house that I like, I've got a nice car standing up in my driveway, I've got money in my account whenever I need to buy food', there's that confidence, that self esteem of wanting to go and learn more and more. And that's what I've come from, and I am really speaking a lot about myself.

6. Housing and infrastructure

At the end of the 1960s and into the 1970s, the migration of Indigenous people off pastoral properties across the Pilbara into emerging urban areas (and consequently away from the inland towards the coast) placed considerable strain on available housing stock in the region and added to the pressures for new dwelling construction (Edmunds 1989: 32). At the 1971 Census, a total of 270 Indigenous dwellings were identified across the Pilbara providing shelter for a total of 2323 residents to produce an average occupancy rate of 8.6 persons per dwelling. Since Indigenous post-censal population estimates were not available at that time, it can only be assumed that this figure represented an undercount of the true occupancy rate.

In the ensuing three decades, difficulties in overcoming the backlog in housing need have been compounded by rapid growth of the Pilbara Indigenous population, as well as increased population dispersion across some 33 or so discrete communities and homeland settlements, 60 per cent of which have fewer than 50 residents. Not surprisingly, in 1991, the first nation-wide normative measure of Indigenous housing need found that the South Hedland and Warburton ATSIC Regional Council areas had relatively high levels of family 'homelessness' and overall housing need (defined for statistical purposes as families in improvised homes, or sharing overcrowded dwellings) in relation to the 36 ATSIC regions nationwide (Jones 1994: 61–4).

The major response to such inadequacies was led by the Commonwealth and developed out of the National Aboriginal Health Strategy (NAHS) in 1990. This recognised an essential link between health outcomes and the provision of housing and infrastructure to acceptable minimum standards. Accordingly, funding allocations in the initial years of the NAHS primary health and environmental health programs included amounts directed at housing and infrastructure services within ATSIC's Community Housing and Infrastructure Program (CHIP). However, a review of CHIP in 1994 identified a range of problems, including a failure to address housing and infrastructure needs in a holistic way. Because of the short-term nature of the program-based approach to funding, communities were being required to structure housing needs to the CHIP program rather than the other way around. A key response to these criticisms was the establishment in 1994 of the Health Infrastructure Priority Projects (HIPP) program to pilot new delivery arrangements for the construction of Indigenous community housing and infrastructure.

In Western Australia, elements of NAHS/HIPP were incorporated into a 1997 bilateral agreement between the State government and ATSIC for the provision of housing and related infrastructure to Indigenous people in the State. This arrangement was updated after review in 2000 with an agreement to pool funding from the Commonwealth, ATSIC and the Western Australian government for the provision of Indigenous housing and infrastructure under the auspices of an Indigenous Housing and Infrastructure Council.

Housing in 2001

The five-yearly census is an enumeration of population and housing. It provides a range of details regarding the number and structure of dwellings and it is possible to classify these according to Indigenous or non-Indigenous occupancy and other housing-related variables. Table 6.1 shows the number, type, and occupancy rate of Indigenous and non-Indigenous dwellings in the Pilbara, with the former classified as such if one or more adults in a dwelling are Indigenous. Because occupancy rates are so directly affected by numbers in each dwelling, the population included in Table 6.1 is the ERP, in an attempt to overcome census undercount and reflect more accurately the adequacy of housing provision, certainly in respect of Indigenous dwellings. The census distribution by dwelling type is simply inflated in line with the ERP.

Table 6.1. Structure of dwellings and occupancy rates for Indigenous and non-Indigenous households: Pilbara SD, 2001

Dwelling type	Indigenous dwellings			Non-Indigenous dwellings		
	Dwellings	Persons[a]	Occupancy rate	Dwellings	Persons[a]	Occupancy rate
Separate house	1007	5426	5.4	6973	26 668	3.8
Town house/apartment	257	853	3.3	1984	4749	2.4
Improvised and other dwellings	56	114	2.0	544	1,270	2.3
Total[b]	1343[b]	6516	4.9	9603[b]	32 947	3.4

[a]ERP-based

[b]Includes structure of dwelling not stated.

Source: ABS 2002b.

In 2001, a total of 10 946 dwelling units were recorded by the census in the Pilbara SD. Of these 1343 (12%) were Indigenous dwellings, most of which were separate houses, with 4 per cent recorded as improvised dwellings (a low proportion when compared to many other parts of remote Australia). While most non-Indigenous dwellings were also separate houses, the key point of distinction is the higher average occupancy rate in Indigenous dwellings – 44 per cent higher than recorded for non-Indigenous dwellings (4.9 compared to 3.4). As a benchmark, it is also interesting to compare this rate with the average of 3.8 persons per Indigenous dwelling recorded for Western Australia as a whole in 2001. Compared to the situation 30 years earlier, however, this is a much lower rate of Indigenous occupancy than the figure of 8.6 recorded by the 1971 Census. While this points to a substantial overall decline in the number of Indigenous persons per dwelling in the Pilbara over the past 30 years, it nonetheless also masks a good deal of internal variation (see Interview segment 40, p. 111, ; Interview segment 42, p. 112).

Indigenous occupancy rates are presented by IA in Table 6.2 p. 99 and ranked from highest to lowest. Once again, these are based on Indigenous ERPs as in Table 2.3. A clear urban/rural divide emerges with all of the remote inland parts of the Pilbara (where a higher proportion of the Indigenous population lives in discrete communities and homeland settlements) reporting occupancy rates well above the Pilbara average, and most of the more urbanised coastal regions reporting below average rates. Thus, the number of persons per dwelling in the Jigalong IA is almost twice the Pilbara average. Against this model, the main exception is Roebourne town which has relatively high occupancy. Thus, 30 years on, there remain locations within the Pilbara where housing occupancy rates appear to have barely altered.

Table 6.2. Indigenous housing occupancy rates by IA: Pilbara SD, 2001

Indigenous Area	Indigenous occupancy rate[a]
Jigalong	8.0
East Pilbara West	7.9
Roebourne (excl. Roebourne town)	7.7
Marble Bar	6.1
Yandeeara	5.8
Port Hedland (ex urban)	5.8
Roebourne	5.4
East Pilbara East	5.4
Ashburton	4.3
Port Hedland	4.0
Karratha	4.0
Pilbara	4.9

[a]Based on ERP.
Source: ABS 2002b.

While the continuance of high Indigenous occupancy rates reflects larger Indigenous household size and a cultural preference for extended family living arrangements, it is also a measure of the inadequacy of housing stock available to accommodate the regional population. To acquire a better sense of the adequacy of housing, occupancy rates must be set against dwelling size, and one measure of this is provided by the ratio of available bedrooms to the population in dwellings. Overall, in the Pilbara, the census recorded a total of 32 331 bedrooms in 2001. Of these, 3715 (11%) were in Indigenous dwellings. Using the number of persons per dwelling inflated to match the ERP, this produces an average figure of 1.8 persons per bedroom in Indigenous dwellings. The equivalent figure for non-Indigenous dwellings is 0.9 – exactly half the Indigenous occupancy rate.

More refined measures also include an indication of housing affordability as well as functionality from an environmental health perspective. Unfortunately these complex calculations are only available at the ATSIC regional level. As far as the population of the Pilbara region is concerned this includes the whole of

South Hedland ATSIC Region, but only the northern part of Warburton ATSIC Region including settlements from Newman eastwards. In order to make use of these data, we assume here that the measures observed for the Warburton Region as a whole apply also to that component located within the Pilbara. Applying basic overcrowding measures, Jones (1994) identified the Warburton ATSIC Region as ranked second, and South Hedland ranked sixteenth highest out of 36 ATSIC Regions across the country in 1991 in terms of the size of their unmet housing need. This was calculated on the basis of additional bedrooms required to meet an accepted occupancy rate. By 2001, Warburton had moved to fourth place while South Hedland remained unchanged at sixteenth (National Centre for Social Applications of GIS 2003: 66).

In terms of the actual number of extra bedrooms required to reduce overcrowding to an acceptable standard level, this actually increased between 1991 and 2001 for both ATSIC regions – from 605 to 731 in Warburton, and from 192 to 504 in South Hedland – as a consequence of population growth in relation to available housing stock (Jones 1994: 55; National Centre for Social Applications of GIS 2003: 66). In 2001, a total of 215 Indigenous households were in overcrowded dwellings in Warburton ATSIC Region, and 180 in South Hedland (National Centre for Social Applications of GIS 2003: 66). As for housing affordability, a total of 24 Indigenous households were paying more than 25 per cent of their household income on rent in 2001 (a relatively low figure due to the high proportion of Indigenous community rental housing), but in South Hedland the equivalent figure was 207 households (National Centre for Social Applications of GIS 2003: 71). In the latter case, one-third (32.4%) of all Indigenous rental households in the lowest two quintiles of household income (482 households) were paying more than 25 per cent of their income as rent (National Centre for Social Applications of GIS 2003: 70).

The depth of this imputed housing need is supported by data from the 2001 CHINS for discrete Indigenous communities across the Pilbara. This reveals an overall occupancy rate for these communities of 7.1 persons per dwelling. However, as Fig. 6.1 indicates, this masks considerable diversity of circumstance, with occupancy rates ranging from 30 persons per dwelling in one instance, to zero in others. Of course, these data reveal nothing of the quality of housing stock.

Figure 6.1. Persons per dwelling at discrete Indigenous communities in the Pilbara, 2001

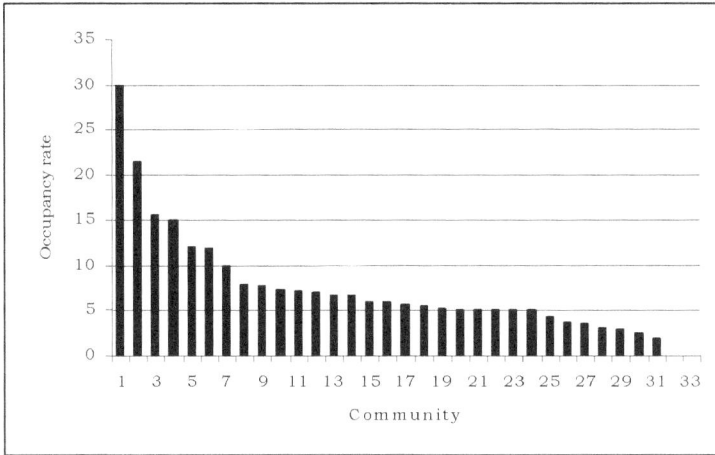

Source: ABS 2001 CHINS.

However, for the first time, the 1997 Western Australia Environmental Health Needs Survey (EHNS) provided for a more refined (and meaningful) measure of occupancy based on persons per functional dwelling (defined against minimum environmental health criteria). This re-calibration produced some excessively high occupancy rates. For example, at Woodstock there were 68 persons per functional dwelling. Overall, communities that were found to have occupancy rates substantially above the regional average, using the stock of functional housing as the base, included Woodstock, Mingullatharndo, Ngalakura, Warralong, Jigalong, Kunawarriji, and Punmu. It is interesting to compare these findings with results from the more recent 2003 EHNS as this later survey suggests a much reduced level of adjusted occupancy levels (Table 6.3).

Table 6.3. Adjusted EHNS population density measures for dwellings in discrete Indigenous communities[a]: Pilbara SD, 2003

Community	Population	Crude PDM	Adjusted PDM
Joongnardhi	15	15.0	15.0
Woodstock	30	7.5	10.0
Cheeditha	58	8.2	9.7
Kunawarritji	110	7.3	8.5
Warralong	50	5.5	7.1
Yandeyarra	320	6.2	6.8
Kiwirrkurra	155	6.2	6.2
Punmu	83	4.9	5.9
Murturakarra	50	5.0	5.6
Bindi Bindi	110	4.1	5.0
Marta Marta	15	5.0	5.0
Youngaleena	30	4.2	5.0
Jinparinya	30	4.2	4.3
Parngurr	104	4.3	4.3
Bellary	25	3.5	4.2
Jigalong	196	4.0	4.2
Tjalka Wara	33	3.3	3.3
Ngurawaana	22	2.2	2.4
Tjalka Boorda	56	2.1	2.2
Pipunya	25	1.6	1.7
Punju Ngamal	7	0.8	0.9

[a]Wakathuni not included.

Source: Government of Western Australia 2003 EHNS.

Further measures of the quality of housing stock are provided by the 2001 CHINS that included an assessment of the condition of dwellings owned or managed by Indigenous Housing Organisations. For the Pilbara, such dwellings were categorised according to the extent of repairs needed in the following way:

- Minor repairs – repairs of less than $33 000
- Major repairs – repairs of between $33 000 and $100 000
- Replacement – repairs of over $100 000

Of the 318 permanent dwellings listed as managed by an Indigenous housing organisation in the Pilbara, all were found to be in need of minor repairs, 131 (41%) needed major repairs, and 25 (8%) required replacement. In effect, according to the 2001 CHINS, a major component of the Indigenous-owned and managed housing stock in the region was in need of significant upgrading, while additional dwellings were also required to reduce overcrowding. If we apply similar grounds as the 1997 EHNS to define functional housing and exclude houses needing major repair or replacement from the available stock, occupancy rates would reach excessively high levels in Cheeditha, Tjalkliwarra, Punmu, Parnngurr, Mugarinya, Wakathuni, Bellary Springs, Strelley, Bindi Bindi, and Youngaleena.

Housing tenure

The Pilbara region is somewhat unusual within the state, indeed within Australia, for having such a high proportion of its private dwellings as company-owned rental housing. In 2001, more than one-quarter of the rental housing stock (27%), and 14 per cent of housing overall was company-owned. This is a legacy (and current reality) of direct corporate investment in the provision of regional infrastructure in order to initiate and support resource development. Initially, of course, a much greater share of the Pilbara housing stock was company-owned as Hamersley Iron constructed the 'closed' towns of Dampier, Tom Price and Paraburdoo, Robe established Pannawonica and Wickham, and Newman Mining invested in the town of Newman. Resource companies also held a major stake in the development of Karratha, Port Hedland and South Hedland.

Since the 1980s, the tenure situation has become more complex as the process of normalising urban tenurial arrangements has gathered pace. However, company ownership of housing remains significant. For example, in 2003, as part of the Pilbara Workforce Delivery Strategy (Hames Sharley 2004: 13–15), the tenure situation with regard to Hamersley Iron was identified as follows: Hamersley Iron owned 531 dwellings in Karratha, of which 293 were available to Hamersley Iron staff and 221 were leased on to the market. Other major corporate home owners in Karratha included Woodside (665 dwellings), as well as Robe and Dampier Salt. In addition to this, a total of 2262 single person transportable temporary accommodation units were identified as required in Karratha for temporary workforces associated with major corporates (Pilbara Development Commission 2004). In Dampier, Hamersley Iron owned 459 out of a total of 575 dwellings; in Tom Price, there were 1372 dwellings with 1038 owned by Hamersley Iron; and they also owned 620 of the 700 dwellings in Paraburdoo. Similar scales of corporate ownership are found in other Pilbara towns such as Newman, Port Hedland and South Hedland.

Home ownership

One consequence of this 'company town' legacy is a relatively low level of home ownership in the Pilbara. Australia as a whole has one of the highest home ownership rates among OECD countries, and in line with this, 67 per cent of all Western Australian households in 2001 lived in a dwelling that was either fully owned or mortgaged. The situation in the Pilbara is outlined in Table 6.4. Less than half (46%) of non-Indigenous dwellings are fully owned or being purchased, and far fewer Indigenous dwellings (14%). Consequently, the predominant form of tenure is rental which accounts for half of all non-Indigenous dwellings, but as much as 80 per cent of Indigenous dwellings. However, the implications of this are likely to be quite different for the two populations. While this limits access to the property market for Indigenous people as a means of improving their financial security, it is also symptomatic of their relatively low economic status as well as some cultural focus on communal forms of tenure. For non-Indigenous people it is far more likely to reflect their preference for investing in housing markets elsewhere (notably Perth) using the proceeds of earnings gained in the Pilbara.

Table 6.4. Indigenous and non-Indigenous dwellings by tenure type:[a] Pilbara SD, 2001

	Fully owned	Being purchased	Being rented	Other tenure type	Total
Indigenous dwellings					
No.	56	131	1028	74	1289
%	4.3	10.2	79.8	5.7	100.0
Non-Indigenous dwellings					
No.	985	3222	4697	520	9424
%	10.5	34.2	49.8	5.5	100.0

[a]Excludes tenure type not stated.
Source: ABS 2002b.

Rental housing

According to 2001 Census data, a total of 5725 dwellings in the region were rented, and Table 6.5 shows the distribution of those for which a landlord type classification was available. Clearly, Indigenous people depend far more on state-provided rental housing than do non-Indigenous people. According to these census data, more than half (55%) of Indigenous dwellings in the Pilbara are rented from the Western Australia Department of Housing and Works (DHW). This compares to only 11 per cent of non-Indigenous rental dwellings. The main reason for this contrast is the much greater access to employer-owned accommodation for non-Indigenous workers, especially in the non-government sector (see Interview segment 35, p. 109; Interview segment 36, p. 110). Almost one-third of non-Indigenous rental dwellings are provided by non-government employers. Also of note is the greater use of private rental among non-Indigenous households, with one-third renting from a private landlord or real estate agent compared to just 11 per cent of Indigenous households. Once again, this reflects the temporary nature of many non-Indigenous households, but it also reflects relative housing affordability judging by median household incomes shown in Table 4.1. Partly for this same reason, Indigenous households are largely restricted in their housing options to the state rental sector, while in discrete communities Indigenous community housing association dwellings predominate (see Interview segment 35, p. 109; Interview segment 36, p. 110; Interview segment 38, p. 111).

Table 6.5. Indigenous and non-Indigenous rental housing by landlord type;[a] Pilbara SD, 2001

	Private landlord	Real estate agent	State housing	Community housing	Govt. Employer	Other Employer	Other landlord	Total
Indigenous dwellings								
No.	37	73	564	203	75	54	16	1022
%	3.6	7.1	55.2	19.9	7.3	5.3	1.6	100.0
Non-Indigenous dwellings								
No.	442	1,057	526	35	1000	1464	134	4658
%	9.5	22.7	11.3	0.8	21.5	31.4	2.9	100.0

[a] Excludes landlord type not stated.

Source: ABS 2002b.

Given the more complex rental options evident in urban settings, it is interesting to compare these census data on rental accommodation with tenancy data for 2004 made available by the DHW (Table 6.6). Basically, DHW funding arrangements provide for three types of urban rental outcomes – mainstream public rental, Indigenous specific rental, and rental dwellings made available to State (or Commonwealth) public servants under the Government Employees Housing Authority (GEHA). As indicated, the DHW currently manages 1489 dwellings across Pilbara towns, a figure which is considerably higher than the 1090 reported by the 2001 Census because it includes elements of other census rental categories, especially government employer housing. As at 31 May, 2005, a total of 1286 public housing dwellings were available in Pilbara towns, a total of 253 dwellings were designated as state owned and managed Indigenous housing, and 1099 GEHA dwellings were available. The distribution of Indigenous tenants in dwellings within the first two of these categories (in 2004) is shown in Table 6.6 using actual DHW tenants data.

Table 6.6. Number of Indigenous and non-Indigenous DHW rental dwellings:[a] Pilbara SD, 2004

	Indigenous (no.)	Non-Indigenous (no.)	Total (no.)	Indigenous (%)
Roebourne	93	6	99	93.9
Karratha	132	322	454	29.1
Wickham	52	28	80	65.0
Onslow	29	8	37	78.4
Tom Price	13	5	18	72.2
Newman	41	18	59	69.5
Paraburdoo	4	1	5	80.0
Marble Bar	21	3	24	87.5
Port Hedland	502	195	697	72.0
Total	887	586	1473	60.0

[a] Includes public housing and Aboriginal-specific public housing. Excludes GEHA and vacant dwellings.

Source: Western Australia Department of Housing and Works.

Overall, 60 per cent of state-owned public housing dwellings are occupied by Indigenous tenants. By far the lowest rate of Indigenous tenancy is found in Karratha, but for the most part Indigenous people are the primary occupants of State-owned public housing, especially in towns such as Roebourne, while Port Hedland also stands out for the sheer number of Indigenous tenancies. However, there are also large numbers of households on waiting lists for housing and these are shown in Table 6.7. Overall, the number of Indigenous households (at least as defined by those applicants for housing who identified to DHW as Indigenous) is 210 which, scale-wise, is almost one quarter (23%) of the total number of Indigenous households already in such housing. However, the number of non-Indigenous households on waiting lists is also relatively large in proportion to the existing number of non-Indigenous occupants (33%). Table 6.7 indicates that the bulk of applications for housing from Indigenous households are in Port Hedland, followed by Newman, whereas for non-Indigenous households, Karratha is the location with the highest demand, followed by Port Hedland. There are two steps in the process of acquiring public housing – first, a management review of the application, and then placement on a waiting list. Table 6.7 reveals that around three-quarters of applicants are wait-listed at any one time, with little significant difference between locations or between Indigenous and non-Indigenous applicants.

Table 6.7. Indigenous and non-Indigenous applications for DHW housing in the Pilbara SD by preferred location and application status, 2005[a]

Preferred location	Indigenous		Non-Indigenous	
	No.	% wait listed	No.	% wait listed
Port Hedland	159	69.8	52	65.4
Marble Bar	5	60.0	3	66.7
Karratha	9	55.6	70	72.9
Roebourne	6	100.0	21	76.2
Newman	22	72.7	19	84.2
Tom Price	6	66.7	4	100.0
Wickham	1	100.0	20	95.0
Paraburdoo	2	100.0	0	0.0
Onslow	0	0.0	7	85.7
Total	210	70.5	196	75.5

[a]As at February 2005.

Source: Western Australia Department of Housing and Works.

Clearly, there is some imprecision in all this, not least because of issues surrounding Indigenous identification in housing records. However, some facts seem assured. First, Indigenous families have fewer urban housing options than others because of their low incomes and relative lack of access to employer housing, either GEHA or company, due to poor labour force and occupational status. Second, their access to urban housing is dependent on continued expansion of the state public housing stock, and a guaranteed major share of this. Third, there is a sizeable unmet need in public housing for both Indigenous and non-Indigenous residents of the Pilbara (see Interview segment 37, p. 110;Interview segment 38, p. 111; Interview segment 39, p. 111), an issue that can also apply to mine workers (see Interview segment 43, p. 112)

Environmental health infrastructure

The idea that Indigenous community housing and infrastructure should be designed, constructed and maintained to support healthy living practices is now firmly embedded in policy following the pioneering work of Pholeros, Rainow and Torzillo (1993) in the Pitjantjatjara Lands. A total of nine such practices are identified, in descending order of priority in terms of impact on health outcomes: capacity to wash people, wash clothes and bedding, remove waste safely, improve nutrition, reduce crowding, separate people from animals, reduce dust, control temperature, and reduce trauma. Each of these refer to different aspects of the functionality of dwellings and their related infrastructure. For example, if the focus is on improving nutritional standards and practices, then 'healthy home hardware' refers to the provision of adequate facilities to store, prepare, and cook food. It also extends to water quality and quantity as a lack of these may lead individuals to purchase bottled water or other beverages, thereby adding to expenditure and increasing reliance on soft drinks and cordials.

The National Indigenous Housing Guide (Commonwealth of Australia 1999) includes a range of detailed design and functionality guidelines to address each of these nine healthy living practices. The key functional area with most guidelines is that involving the supply, usage and removal of water: six of the nine healthy living practices are dependent on these. However, even seemingly obscure health-related housing functions include a wide range of design, maintenance and infrastructural features that require attention (Commonwealth of Australia 1999: 49–57).

As with the measurement of housing need, the status of environmental health infrastructure requires a detailed assessment of functionality and adequacy set against agreed normative criteria. The 2001 CHINS includes information on such issues as water supply, sewerage, drainage and solid waste disposal, but this is more in the form of simply noting the existence or otherwise of infrastructure rather than assessing its functionality and adequacy. Likewise, CHINS data do not allow for the proper assessment of activities related to such issues as dust control, animal health and quality of waterways. For example, with regard to dust control, all that is available from the CHINS is the fact that a certain number of permanent dwellings in communities are on sealed roads. While this provides some indication of the likely extent of dust mitigation as an issue, it is far from adequate as an indicator of progress.

The main, and most recent, source of data regarding the functionality of Indigenous dwelling facilities in discrete Indigenous communities in the Pilbara is the Western Australian Government's 2003 EHNS. Unfortunately output on the functionality of dwelling facilities is reported only for all communities collectively within each ATSIC Region. Thus, we have data for all discrete communities in the Ngarda Ngarli Yarndu ATSIC Region, and for the Warburton ATSIC Region, but as only a fraction of the communities within the latter region fall within the Pilbara SD, results are presented here for the Ngarda Ngarli Yarndu ATSIC Region only.

In this region, 15 per cent of dwellings had no external sanitary plumbing and as many as 54 per cent had no on-site sewerage disposal (Table 6.8). The other absent key dwelling facilities related mostly to temperature control with no ceiling insulation, heating, or air conditioning in 29 per cent, 82 per cent and 57 per cent of all dwellings respectively. Even where dwelling facilities were present, more than 20 per cent of showers and toilets were not working, while between 10 and 16 per cent of hot water systems and laundry facilities were also defective. Overall, 35 per cent of dwellings were either without facilities or had facilities that were not working. Among the communities that stood out as having more than 10 per cent of their dwellings with more than three facilities not working were Pipunya, Tjalka Wara, Youngaleena, Yandeyarra, Jinparinya, Warralong, and Murturakarra, with by far the highest percentages of defective dwellings in the first two, although problems with housing functionality are clearly more widespread (see Interview segment 35, p. 00; Interview segment 40, p. 00). In the Warburton ATSIC Region, of the localities falling within the Pilbara SD, only Kunawarritji had more than 10 per cent of its dwellings with more than three defective facilities.

Table 6.8. Functionality of dwelling facilities in Ngarda Ngarli Yarndu ATSIC Region, 2003

Functionality of dwelling facilities	Dwellings with facility absent		Dwellings with facility working		Dwellings with facility not working		Total dwellings surveyed
	(no.)	%	(no.	(%)	(no.)	(%)	(no.)
External sanitary plumbing	28	14.6	149	77.6	15	7.8	192
On-site sewerage disposal	103	53.6	81	42.2	8	4.2	192
Hot water system	8	4.7	140	81.4	24	14.0	172
Kitchen sink	5	2.9	150	87.2	17	9.9	172
Bath and/or shower	5	2.9	120	69.0	49	28.2	174
Toilet cistern	6	3.4	122	68.9	49	27.7	177
Toilet bowl	6	3.3	136	75.6	38	21.1	180
Laundry trough	9	5.1	138	78.9	28	16.0	175
Laundry floor waste outlet	24	13.7	132	75.4	19	10.9	175
Ceiling insulation	54	28.7	128	68.1	6	3.2	188
Heating	158	81.9	31	16.1	4	2.1	193
Air conditioning	108	57.4	75	39.9	5	2.7	188
Ceiling fan	20	10.8	128	69.2	37	20.0	185

Source: Government of Western Australia 2003 EHNS.

As a summary device, the EHNS provides information on the number of communities reporting particular identified infrastructure needs. Because these are grouped for all communities in each ATSIC Region, only information for the Ngarda Ngarli Yarndu ATSIC Region is presented here (Table 6.9). Thus, out of the 17 communities surveyed in this region, 41 per cent identified housing and municipal services as key needs, 35 per cent identified health hardware, water power and sewerage, 29 per cent identified recreational facilities, 23 per cent identified plant and vehicle workshop equipment, and 18 per cent identified environmental programs such as dust control.

Table 6.9. Identified infrastructure needs: discrete Indigenous communities in the Ngarda Ngarli Yarndu ATSIC Region, 2003

Identified need	No. of communities	% of communities
Housing (new, repairs, housing for visitors and workers)	7	41.2
Municipal Services (street lighting, rubbish disposal, drainage)	7	41.2
Health Hardware (ablutions, hot water systems, washing machines)	6	35.3
Health Services (medical centre, detox centres, first aid kit)	6	35.3
Water, Power, Sewerage (improvements or provision)	6	35.3
Recreational facilities (sporting grounds, play grounds)	5	29.4
Plant/Vehicle Workshop (tools, machinery, tractors, equipment)	4	23.5
Environmental Programs (greening, dust suppression)	3	17.6
Access (internal and access roads, vehicles, boats, airstrips, fuel)	2	11.8
Fencing (houses, tips, sewerage ponds)	2	11.8
Meeting Areas (administration facilities, general purpose buildings)	1	5.9
Telecommunications (phones)	1	5.9
Training (employment and business development)	1	5.9

Source: Government of Western Australia 2003 EHNS.

Of course, these data refer only to dwellings in select discrete Indigenous communities and mostly to housing stock managed by Indigenous housing organisations. No information is available on the quality of other rental accommodation, although Indigenous people do raise issues about this (see Interview segment 35, p. 00).

Indigenous perspectives

Interview segment 35

There are hardly any Homeswest houses here in Tom Price, it's all for mining, and we got to go through real estate to get accommodation. Where are people going to get money for bond? They want bond and two weeks rent in advance. Then you have to fit their criteria. Most of the houses in Tom Price are private rentals. I got my two daughters here but they need another house because they got kids too you see. There are no houses when kids grow up and have their own kids and they have to live with us. My kids they don't want to live like that, they want to have their own house. We also get a lot of family visitors, but Aboriginal people are close-knit families, we don't tell family to go away.

My house isn't really a good house, it's got no air-conditioning, I have to buy my own, and it's small for a three bedroom house. You notice round here that all the houses that Aboriginal people get are small but have a look at the mining houses, they get carport, verandah all around, and shed. When people come to visit me they have to chuck their swag on the porch.

With Aborigine people, if they miss a rent they'll be on their back straight away. But when they want repairs or anything to be done they take months and months maybe even 12 months to start doing stuff you know, even if it's emergency stuff, they won't jump. You gotta put up front money, and I say, 'I'm willing to pay for it, if you start to upgrade my house'. We had that thing with the water bill, these taps are buggered and we had a leak, and Homeswest can't even come and fix the washer or whatever is wrong and that's where the water bill is going upper and upper. No wonder our water bills get high.

Another thing again, we had a busted pipe, and anyway water was coming through the bedroom now, and also through the kitchen and one of the relations came around and had a look for me and said you got busted pipe, they had to make a hole in the wall to find where the water was coming from. There is no main office here where the people can go to report anything urgent. We had to go through Hedland. Centrelink got a sub office here, it wouldn't hurt for Homeswest to do the same eh? You ring up to Hedland and they tell you, you got to wait for them people when you get burst water pipe after-hours. So every night I gotta go to the street tap to turn my water off so I don't have to hear the water running when I go to sleep. We gotta put all the drinking water in bottles. This been happening for two months. That's bad, that's really bad. We just want them to fix the houses up so we can stay in good houses.

Interview segment 36

There are not enough houses in Tom Price. Aboriginal people are stuck up in one house. They can't get another house, but the company building more houses for the boom coming up, more houses for them. But the workers gonna fly-in-fly-out, what about the bloomin' local all stuck up in one house? We can't even get a Homeswest house in town. Some out on blocks, but even they have the same problem out on the block. They not fixing up the community and people then come to town to find a roof.

Interview segment 37

A big problem is not enough housing. When kids grow up and get married, start having their own families, they got nowhere to move. So they end up all in one house and overcrowding. Sure they could move somewhere else, but then they have to leave family. A lotta family arguments start like that too, if one not putting in enough money for rent or bills or whatever. And that stress starts with some and they start drinkin'. Sometimes hard to get a good night's sleep too, which makes it hard for kids going to school you know?

Interview segment 38

People in Tom Price feel neglected being in an isolated place and not having the same benefits as the growth area, which would be Karratha and Port Hedland. And lack of housing here because you find that you have to be working for a mining company to obtain housing here. I'm in that situation because I am only doing private work on my family thing, but I can't get housing here because the amount of rent you have to pay is above my level of income. So I'm the same I get disheartened and want to pack up and leave.

Because there's not enough housing that's when you get the pile in and that's where stress is the biggest problem of all because you get the arguments in there and the shares of your financial income to go towards food, electricity, water, oh yeah, and that's where your family rifts are caused, you know family arguments … and they turn to alcohol just to shut off from it, or the drugs, and that is an issue, it's a big issue. We also understand on the mining part of it that we can't go out and build anything until our native title is all complete. Land access is one of the biggest issues otherwise we could go and build about 50 Homeswest houses out there at Bellary or Wakathuni, but because of the restrictions of land issues, and you know that's all red tape stuff with the politics and the Native Title Act restricts that.

Interview segment 39

Housing has always been a big problem across the board. Very much so for Aboriginal people, because you got to use Homeswest. The waiting lists are long. The locals found that other people coming from out of town were getting housed sooner. We found Homeswest were giving people who may have been on waiting lists in other places in Roebourne houses. People don't want to move out of Roebourne, but since mining decreased a bit a while ago, Homeswest acquired houses in Wickham which used to be a mining town. The housing requirement for miners has decreased in this area.

Interview segment 40

We are all focusing on the one aim which is to bring our oldies and our children back to our homeland, educate them, live on the land so they got more experience and learning because we take em out see? And we'll teach them and they will listen to how we gotta talk to the mining company and anthropologist and that's their learning process. And that's how we're proceeding and that's how we have been all these years. We got 30 people living in one house. I called to one of the housing mob. We got no access to funding or health things and no one servicing us for rubbish or water or anything. Sometimes they come out to check the water quality. A lot of money gets spent on surveys and studies but nothing ever gets followed up.

Interview segment 41

I got kicked out of my Homeswest house. I been go away for three days and them kids been go right through there and breaking thing and making hole everywhere. They broke in through that aircondition window, that's how they got through. Windows broken, doors smashed up everywhere. I went and reported it to police and Homeswest when I come back. Then they sent me a bill for $3000. They still wanted me to pay for it, even though it wasn't me. I lost a lot of furniture too, and I can't get anything for that. I got nothing. They take $100 everytime pay week, and I only get $400 each pay. I got no house. Stay here with family until I pay for damage. Then they give me house back. I gotta look after my son too. He no good in the head. He was a sniffer and now I gotta look after him all the time.

Interview segment 42

Well it happens to old people all the time, they get kicked out. Young fellas come round when no one looking and break everything. They starving all the time, or looking for blanket or something. My mum she's cripple, in a wheelchair you know, and them young fellas broke her door when she was out at a meeting. Well she got kicked out, and got a big debt too. Housing is a real problem. Young people boss the old people around trying to get a swag and smoke and it's very bad. Young people got no home. They like the birds, they got no home, just wander around. They walk camp to camp. Drinking is a big problem for young people in Hedland. I've got four old people living at my house because they got kicked out like that. They can't get a house until they pay that debt. Some of them will pay that debt for the rest of their lives. We also get a lot of family visitors staying. We got 16 people living here in this house. It's OK, as long as everyone buys a little bit of tucker for the fridge.

Interview segment 43

I know a lot of young Banyjima fellas who belong up here, marrying Ngarluma Yindjibarndi girls and they don't like living out here. The wives of the men would rather they stay and work up there on the coast. Fly-in-fly-out is very hard way to work. Ask anyone with wife, and kid, it's very hard. Some locals bring their family with them, but sometime you can only get single quarters. I was here for six months before my family came because I could only get single quarters.

7. Health status

A primary barrier to the enhanced participation of Indigenous people in the Pilbara labour market is poor health status and associated high mortality. According to the Epidemiology Branch of the Western Australia Department of Health, life expectancy at birth for Indigenous males is just 55 years in the East Pilbara Health District (Port Hedland and East Pilbara SLAs), and 52 years in the West Pilbara (Roebourne and Ashburton SLAs). The equivalent figures for females are 60 and 63 years respectively (Pilbara Population Health Unit 2004: 5–21). According to the same source, the Indigenous population accounted for an average of 12 per cent of the West Pilbara population between 1997 and 2001, but for 44 per cent of all deaths. The equivalent figures in the East Pilbara were 19 and 42 per cent (Pilbara Population Health Unit 2004: 10-111-14). Despite this, self-perceptions of health status and health services can often be positive (see Interview segment 45, p. 127; Interview segment 47, p. 127; Interview segment 48, p. 127).

By these facts alone, the chances of full and prolonged Indigenous participation in the workforce are clearly curtailed. For example, using national life tables, the chances that a newborn Indigenous male will reach workforce age (15 years) have been estimated at 97 per cent (Kinfu & Taylor 2002). For those who do reach workforce age, 19 per cent will be dead by the age of 45, and 25 per cent will not reach 50 years of age. Statistically, more than half of Indigenous males who reach age 15 have no chance of surviving to retirement age at 65 years. Thus, out of an average cohort of 100 Indigenous males aged 15, only 48 would still be alive by age 65 (Kinfu & Taylor 2002: 10). Similar, if not lower, survival probabilities apply in the Pilbara. Equally telling, though, is the morbidity profile that underpins this high mortality. Here we observe the cumulative impact of progressive morbidity that can commence often prior to birth, persist through childhood, and become compounded in adult years. Allied to this are lifestyle factors associated with overcrowded dwellings, risk behaviour, low incomes, and poor nutrition. All this is well documented (ABS & Australian Institute of Health and Welfare (AIHW) 2003; Zubrick et al. 2004) and confirms the importance of social and economic determinants of Indigenous health outcomes.

Information on the health status of Indigenous people is collected as a matter of course in the day-to-day operation of the health care system in Western Australia. For the Pilbara, much of that available from the government-run system has been brought together in summary form by the Pilbara Population Health Unit (2004) and this provides a firm basis for establishing the relative health status of the region's Indigenous population. It also allows for a regional disaggregation of health status between the East and West Pilbara Health Districts, although the general finding is one of shared characteristics between these regions.

Mortality

Between 1994 and 2003 a total of 470 deaths were recorded among Indigenous residents of the Pilbara SD. Of these, 271 (58%) were male, and 199 (42%) were female. This number of deaths in the Indigenous population was between 6 per cent and 28 per cent higher than expected when compared to the Indigenous mortality rate for Western Australia as a whole (Western Australia Department of Health 2005). As for a comparison with the rest of the population, the Indigenous age-standardised mortality rate for the Pilbara was more than three times higher than the crude death rate than observed in the Australian population as a whole (23.5 deaths per 1000 compared to 6.4) (Western Australia Department of Health 2005).

Cause of death

Cause of death data are coded using the World Health Organisation (WHO) method of disease classification that follows the 9th Revision, International Classification of Diseases (ICD9) up to July 1999, and the ICD10 classification thereafter. In Western Australia as a whole, the highest rates of Indigenous deaths are seen in cancer, diseases of the circulatory system, respiratory diseases, endocrine disorders (especially diabetes) and injury and poisoning. During the 1990s, these disease categories accounted for 75 per cent of all Indigenous deaths in the state (Watson, Ejueyitsi & Codde 2001). As indicated in Table 7.1, this is the same proportion observed for all Indigenous male deaths in the Pilbara between 1994 and 2003, although these categories accounted for only 66 per cent of Indigenous female deaths over the same period.

Table 7.1. Indigenous deaths by cause (ICD9): Pilbara SD, 1994–2003

ICD 9 disease chapter	Deaths		Proportion of deaths	
	Males (no.)	Females (no.)	Males (%)	Females (%)
Infectious and parasitic	5	5	1.8	2.5
Cancer	33	26	12.2	13.1
Endocrine/nutritional	10	26	3.7	13.1
Blood diseases	0	0	0.0	0.0
Mental disorders	12	9	4.4	4.5
Nervous system diseases	4	5	1.5	2.5
Circulatory diseases	78	48	28.8	24.1
Respiratory diseases	26	15	9.6	7.5
Digestive diseases	17	17	6.3	8.5
Genitourinary diseases	6	9	2.2	4.5
Complication pregnancy	0	0	0.0	0.0
Skin diseases	1	1	0.4	0.5
Musculoskeletal diseases	2	1	0.7	0.5
Congenital anomalies	6	3	2.2	1.5
Perinatal conditions	3	6	1.1	3.0
Ill-defined conditions	11	11	4.1	5.5
Injury and poisoning	57	17	21.0	8.5
All causes	271	199	100.0	100.0

Source: Western Australia Department of Health 2005.

Age-standardised death rates for each of these five main conditions have been calculated by the Western Australia Department of Health for the East and West Pilbara, with comparison drawn between Indigenus and non-Indigenous residents. These results are shown in Table 7.2 and 7.3, together with the Indigenous to non-Indigenous rate ratios. Thus, in the East Pilbara, between 1990 and 1999 there were 102 Indigenous deaths due to circulatory disease and this produced an age-standardised rate that was 3.9 times greater than for non-Indigenous residents. As indicated, this difference was statistically significant. Indigenous death rates were also significantly higher than non-Indigenous rates for respiratory disease and injury and poisoning. Deaths rates due to cancer were not significantly different, while diabetes deaths were too few to draw meaningful conclusions. Essentially the same pattern was observed in the West Pilbara.

Table 7.2. Indigenous and non-Indigenous age-standardised mortality rates for selected major health conditions for East Pilbara, 1990–1999

Condition	Indigenous		Non-Indigenous		Rate ratio
	No.	ASR	No.	ASR	
Circulatory Disease	102	647.5	68	164.6	3.9 [H]
Cancer	42	266.7	83	162.4	1.6 [NS]
Respiratory Disease	37	218.5	18	50.4	4.3 [H]
Injury and Poisoning	55	208.0	73	46.7	4.5 [H]
Diabetes	13	91.1	7	25.2	3.6 [#]

Key: [H] = Significantly higher than the non-Indigenous population in the region; [NS] = Not significantly different from the non-Indigenous population in the region; [#] = Number of cases too low to draw meaningful conclusions.
Source: Watson, Ejueyitsi and Codde 2001.

Standardised mortality rate ratios were also calculated for each of the Pilbara health districts to test for any significant internal variation in Indigenous and non-Indigenous rates for each of these conditions. As all of these ratios were very close to 1.0 no significant spatial variation in mortality rates within the Pilbara was evident (Pilbara Population Health Unit 2004: 9-94-96).

Table 7.3. Indigenous and non-Indigenous age-standardised mortality rates for selected major health conditions for West Pilbara, 1990–1999

Condition	Indigenous		Non-Indigenous		Rate ratio[a]
	No.	ASR	No.	ASR	
Circulatory Disease	34	538.6	46	122.9	4.4 [H]
Cancer	18	266.5	64	109.7	2.4 [NS]
Respiratory Disease	13	262.5	10	33.2	7.9
Injury and Poisoning	20	195.6	54	31.4	6.2 [H]
Diabetes	7	87.2	4	11.4	7.6[#]

[a]Age-standardised with the Australian population and expressed per 100 000 population.
Key: [H] = Significantly higher than the non-Indigenous population in the region; [NS] = Not significantly higher than the non-Indigenous population in the region; [#] = Number of cases too low to draw a meaningful conclusion.
Source: Watson, Ejueyitsi, and Codde 2001.

In Tables 7.4 to 7.7, the top 15 causes of death are examined in more detail for Indigenous males and females in the East and West Pilbara Health Districts. Also shown is the standardised mortality rate ratio (SRR) using the state Indigenous data as the standard. This provides an indication of the relative significance of rates observed in the Pilbara regions compared to the state Indigenous rates for males and females. Confidence intervals (CIs) wholly above or wholly below 1.0 (the state rate) indicate that rates in the Pilbara are significantly higher or lower. Quick perusal of these CIs reveals that none of the rates for the detailed top 15 causes of mortality for Indigenous males and females in the East or West Pilbara is significantly different from the equivalent rates observed at the State level. In short, from a Western Australian perspective, there appears to be nothing unusual about the profile of Indigenous mortality throughout the Pilbara.

Thus, as elsewhere in the state, ischaemic heart disease, cerebrovascular disease, transport related accidents, and non-specified cancers are all prominent among the major causes of death for Indigenous males, while for females diabetes, liver disease and other forms of heart disease are also prevalent. This profile of mortality confirms the trend towards 'lifestyle' diseases as the primary cause of death in remote Western Australia first noted by Gracey and Spargo (1987) in their review of the state of Indigenous health in the Kimberley region for the period 1970 to 1985. At the same time, specific mining-related causes are also reported. For example, all known cases of malignant mesothelioma among Indigenous people in Western Australia are reported from the Pilbara (Musk et al. 1995). Most of these cases are due to exposure while involved in the transport of asbestos from the Wittenoom crocidolite operation. According to Musk et al. (2005) the incidence of this disease among Indigenous people in the Pilbara (250 per million for ages 15 and over) is one of the highest population-based rates recorded anywhere in the world, with the likelihood that the risk of mesothelioma resulting from past exposures will continue to rise over time.

Table 7.4. Top 15 causes of mortality for the Indigenous male population of the West Pilbara Health District, 1993–2002

Condition	no.	% of all cases	SRR	Confidence interval
Ischaemic heart disease	17	17.5	1.2	0.65–1.86
All other cancers	12	12.4	1.8	0.84–3.02
Chronic obstructive pulmonary disease	5	5.2	1.8	0.40–3.74
Diabetes	5	5.2	1.1	0.23–2.15
Cerebrovascular disease	5	5.2	0.9	0.21–1.93
Suicide and self inflicted injury	5	5.2	0.9	0.19–1.74
Transport related accidents	4	4.1	0.6	0.09–1.33
Neurotic, personality, mental disorders	4	4.1	1.8	0.27–3.88
Other infectious & parasitic diseases	4	4.1	3.6	0.55–7.84
Liver diseases	3	3.1	0.9	0.07–1.97
Ill-defined causes	3	3.1	0.8	0.07–1.95
Pneumonia and influenza	3	3.1	0.8	0.06–1.74
Congenital anomalies	3	3.1	2.2	0.18–5.20
Accidental poisoning	2	2.1	1.2	0.01–3.23
Organic psychotic conditions	2	2.1	2.0	0.03–5.58

Source: Pilbara Population Health Unit 2004.

Table 7.5. Top 15 causes of mortality for the Indigenous female population of the West Pilbara Health District, 1993–2002

Condition	no.	% of all cases	SRR	Confidence interval
Diabetes	8	14.5	1.1	0.40–2.03
Ill-defined causes	5	9.1	2.6	0.56–5.28
Liver diseases	5	9.1	2.1	0.46–4.36
Ischaemic heart disease	4	7.3	0.5	0.08–1.14
Transport related accidents	3	5.5	1.1	0.09–2.54
Cerebrovascular disease	3	5.5	0.7	0.06–1.71
Pneumonia and influenza	2	3.6	0.9	0.01–2.37
Other forms of heart disease	2	3.6	0.8	0.01–2.20
Other diseases of digestive system	2	3.6	7.1	0.09–19.65
Neurotic, personality, mental disorders	2	3.6	2.6	0.03–7.19
Congenital anomalies	2	3.6	2.3	0.03–6.27
Lung cancer	2	3.6	1.5	0.02–4.27
Other conditions in perinatal period	2	3.6	2.4	0.03–6.75
Accidents caused by submersion, suffocation & foreign bodies	2	3.6	2.8	0.04–7.89
Other disorders of CNS	1	1.8	1.2	0.00–4.48

Source: Pilbara Population Health Unit 2004.

Table 7.6. Top 15 causes of mortality for the Indigenous male population of the East Pilbara Health District, 1993–2002

Condition	no.	% of all cases	SRR	Confidence interval
Ischaemic heart disease	23	12.8	0.93	0.55–1.34
Cerebrovascular disease	17	9.5	1.62	0.87–2.47
Transport-related accidents	16	8.9	1.88	0.99–2.91
All other cancers	13	7.3	1.1	0.52–1.77
Pneumonia and influenza	9	5.0	1.27	0.49–2.22
Other forms of heart disease	7	3.9	1.21	0.38–2.26
Chronic obstructive pulmonary disease	6	3.4	1.14	0.31–2.22
Diabetes	6	3.4	0.7	0.19–1.36
Ill-defined causes	6	3.4	1.15	0.31–2.23
Liver diseases	6	3.4	1.1	0.30–2.14
Other diseases of digestive system	6	3.4	3.54	0.96–6.89
Lung cancer	5	2.8	1.43	0.31–2.93
Other diseases of respiratory system	5	2.8	2.49	0.54–5.09
Suicide and self inflicted injury	5	2.8	0.77	0.17–1.58
Other accidents	4	2.2	1.74	0.27–3.81

Source: Pilbara Population Health Unit 2004.

Table 7.7. Top 15 causes of mortality for the Indigenous female population of the East Pilbara Health District, 1993–2002

Condition	no.	% of all cases	SRR	Confidence interval
Ischaemic heart disease	18	11.0	1.12	0.62–1.70
All other cancers	16	9.8	1.89	0.99–2.92
Cerebrovascular disease	15	9.2	1.61	0.82–2.52
Diabetes	13	8.0	0.86	0.41–1.38
Other forms of heart disease	10	6.1	1.84	0.76–3.14
Liver diseases	9	5.5	2.4	0.92–4.19
Transport-related accidents	7	4.3	1.62	0.51–3.02
Neurotic, personality, mental disorders	6	3.7	4.35	1.18–8.46
Ill-defined causes	6	3.7	1.71	0.46–3.33
Pneumonia and influenza	6	3.7	1.29	0.35–2.51
Accidents caused by submersion, suffocation & foreign bodies	4	2.5	3.23	0.50–7.09
Organic psychotic conditions	4	2.5	1.89	0.29–4.15
Other infectious & parasitic diseases	4	2.5	3	0.46–6.57
Nephritis, nephrotic syndrome & nephrosis	4	2.5	3.67	0.57–8.04
Renal failure (acute & chronic)	4	2.5	1.67	0.26–3.67

Source: Pilbara Population Health Unit 2004.

Morbidity

Hospital separations data for the Indigenous and non-Indigenous usual resident populations of the East and West Pilbara Health Districts have been compiled by the Pilbara Population Health Unit for the years 1998–2002. These data form the basis for compiling a statistical profile of the health status of the regional population. However, because the focus is inevitably on diagnoses of major morbidity (i.e. conditions serious enough to warrant hospitalisation), they do not provide a full measure of the burden of ill health in the region. For this we would need to add indicators of health status from primary health care providers.

Before considering hospitalisation data in detail, it is important to note that the number of admissions far exceeds the number of individuals admitted. This is obviously because many people are admitted more than once. Although unique patient data are not reported by the Population Health Unit, previous analysis from the East Kimberley (Taylor 2004a) suggests that an average of 1.9 separations per Indigenous patient compared to 1.02 separations per non-Indigenous patient might apply, making the Indigenous hospitalisation ratio twice as high.

Between 1998 and 2002, a total of 32 162 hospital separations were recorded among residents of the East Pilbara Health District population (Pilbara Population Health Unit 2004: 11–127). Of these, 14 731 (46%) were Indigenous separations, even though Indigenous people represented just 19 per cent of the average sub-regional population. In the West Pilbara, the overall number of separations was fewer at 25 142, and Indigenous people accounted for 6755 (27%) of these despite comprising 12 per cent of the Health District population (Pilbara Population Health Unit 2004: 11–12). In both sub-regions, then, the level of Indigenous hospitalisation was more than double that suggested by their population share.

Hospitalisation diagnoses

In profiling the nature of morbidity as defined by principal disease diagnoses, data for all hospital separations (including repeat separations) are utilised. This is because individuals can, and often are, admitted to hospital more than once, but for quite different reasons. Tables 7.8–7.11 show the distribution of the top 15 causes of hospitalisation for Indigenous males and females separately in the East and West Pilbara Health Districts over the period 1994–2001.

Table 7.8. Top 15 causes of hospitalisation for the Indigenous male population of the West Pilbara Health District, 1994–2001

Condition	no.	% of all cases	SRR	Confidence interval
Other injuries^	641	10.5	1.23	1.13–1.32*
Encounter for dialysis^	593	9.7	0.53	0.49–0.58#
Ill-defined causes	522	8.5	1.82	1.67–1.99*
Diabetes^	479	7.8	5.95	5.44–6.52*
Pneumonia and influenza^	436	7.1	2	1.82–2.20*
Acute respiratory infections^	350	5.7	2.22	2.00–2.47*
Infections of skin & subcutaneous tissue^	247	4.0	1.32	1.16–1.50*
Fractures and sprains	204	3.3	1.03	0.89–1.18
Other factors influencing health and contact with service	192	3.1	1.52	1.32–1.76*
Other disorders of CNS	167	2.7	1.09	0.94–1.28
Bronchitis (acute and chronic)^	162	2.6	4.67	3.99–5.46*
Disorders of the ear^	154	2.5	2.06	1.76–2.42*
Other diseases of respiratory system+	145	2.4	1.42	1.20–1.68*
Neurotic, personality, mental disorders	141	2.3	0.9	0.76–1.06
Intestinal infectious diseases^	135	2.2	1.9	1.60–2.25*

Key: ^ = Showed a significant decrease in the West Pilbara population over the 5 year period 1997–2001; + = Showed a significant increase in the West Pilbara population over the 5 year period 1997–2001; * = Compared to the state Indigenous rate, the number of hospitalisations was significantly greater than expected; # = Compared to the state Indigenous rate, the number of hospitalisations was significantly lower than expected.
Source: Pilbara Population Health Unit 2004.

Table 7.9. Top 15 causes of hospitalisation for the Indigenous female population of the West Pilbara Health District, 1994–2001

Condition	no.	% of all cases	SRR	Confidence interval
Other injuries	645	10.8	1.43	1.32–1.55*
Ill-defined causes	398	6.6	1.39	1.26–1.54*
Pneumonia and influenza^	349	5.8	2.2	1.98–2.45*
Acute respiratory infections^	319	5.3	2.44	2.18–2.72*
Complication related to pregnancy^	317	5.3	0.86	0.77–0.96#
Encounter for dialysis^	236	3.9	0.13	0.12–0.15#
Infections of skin & subcutaneous tissue^	231	3.9	1.56	1.37–1.78*
Other diseases of urinary system^	220	3.7	1.83	1.60–2.09*
Other factors influencing health and contact with service	202	3.4	2	1.74–2.30*
Bronchitis (acute and chronic)	198	3.3	5.19	4.51–5.98*
Asthma	190	3.2	1.41	1.22–1.62*
Fractures and sprains	183	3.1	1.52	1.32–1.77*
Normal delivery and other indications for care	158	2.6	0.68	0.58–0.79#
Other diseases of respiratory system	142	2.4	1.48	1.25-1.75*
Disorders of the ear	135	2.3	1.95	1.64-2.32*

Key: ^ = Showed a significant decrease in the West Pilbara population over the 5 year period 1997–2001; * = Compared to the state Indigenous rate, the number of hospitalisations was significantly greater than expected; # = Compared to the state Indigenous rate, the number of hospitalisations was significantly lower than expected.
Source: Pilbara Population Health Unit 2004

Table 7.10. Top 15 causes of hospitalisation for the Indigenous male population of the East Pilbara Health District, 1994–2001

Condition	no.	% of all cases	SRR	Confidence interval
Encounter for dialysis +	1373	15.9	0.84	0.79–0.88#
Other injuries	824	9.5	91.25	1.16–1.33*
Ill-defined causes	631	7.3	1.54	1.42–1.66*
Pneumonia and influenza	562	6.5	1.79	1.65–1.95*
Infections of skin & subcutaneous tissue	473	5.5	1.89	1.73–2.07*
Fractures and sprains	333	3.9	1.35	1.21–1.51*
Acute respiratory infections^	323	3.7	1.4	1.25–1.56*
Intestinal infectious diseases^	264	3.1	2.54	2.24–2.87*
Other diseases of digestive system +	242	2.8	1.97	1.73–2.24*
Other diseases of respiratory system^	232	2.7	1.5	1.32–1.71*
Other factors influencing health and contact with service	192	2.2	1.03	0.89–1.19
Other disorders of CNS	162	1.9	0.78	0.67–0.92#
Asthma^	157	1.8	1.15	0.98–1.35
Disorders of eye	156	1.8	1.79	1.53–2.10*
Other forms of heart disease	155	1.8	1.39	1.18–1.63*

Key: ^ = Showed a significant decrease in the East Pilbara population over the 5 year period 1997–2001; ' = Showed a significant increase in the East Pilbara population over the 5 year period 1997–2001; * =

Compared to the state Indigenous rate, the number of hospitalisations was significantly greater than expected; # = Compared to the state Indigenous rate, the number of hospitalisations was significantly lower than expected.
Source: Pilbara Population Health Unit 2004.

Table 7.11. Top 15 causes of hospitalisation for the Indigenous female population of the East Pilbara Health District, 1994–2001

Condition	no.	% of all cases	SRR	Confidence interval
Encounter for dialysis+	3344	24.8	1.16	1.12–1.20*
Other injuries	1277	9.5	1.8	1.70–1.90*
Ill-defined causes	674	5.0	1.42	1.32–1.54*
Complication related to pregnancy^	550	4.1	0.96	0.89–1.05
Pneumonia and influenza	459	3.4	1.7	1.55–1.87*
Infections of skin & subcutaneous tissue	452	3.4	1.85	1.69–2.03*
Normal delivery	432	3.2	1.2	1.09–1.32*
Fractures and sprains	363	2.7	1.88	1.70–2.09*
Complication in course of labour	279	2.1	1.05	0.93–1.18
Other pregnancy with abortive outcome	268	2.0	1.39	1.23–1.57*
Acute respiratory infections^	266	2.0	1.17	1.04–1.32*
Other diseases of urinary system	261	1.9	1.26	1.12–1.43*
Other diseases of respiratory system^	249	1.8	1.49	1.31–1.69*
Asthma^	249	1.8	1.08	0.95–1.23
Other factors influencing health and contact with service	239	1.8	1.35	1.19–1.54*

Key: ^ = Showed a significant decrease in the East Pilbara population over the 5 year period 1997–2001; + = Showed a significant increase in the East Pilbara population over the 5 year period 1997–2001; * = Compared to the state Indigenous rate, the number of hospitalisations was significantly greater than expected; # = Compared to the state Indigenous rate, the number of hospitalisations was significantly lower than expected.
Source: Pilbara Population Health Unit 2004.

These data highlight emphatically that chronic diseases (cardiovascular, cancer, chronic pulmonary, and diabetes) have become the dominant causes of morbidity and mortality in the Pilbara, as indeed they have for Indigenous people across Australia (ABS & AIHW 2003: 129–51, 193). Of these, the disease that presents the highest (and growing) gap in rates between Indigenous and non-Indigenous populations is diabetes (especially in the West Pilbara). This is a significant observation given that diabetes is a debilitating condition with several co-morbidities including obesity, high blood pressure, high cholesterol, peripheral neuropathy, blindness, and renal disease. From the individual's perspective, and that of the health care system, it involves a high management regime and is costly in terms of management time and resources. For example, a recent paper cited in Rowbottom et al. (2003: 3) estimates the cost to the Australian health system per year of one diabetic person with complications at over $9000. In addition, related government subsidies on pensions and sickness benefits amount to more than $6000 per year. As far as individuals are concerned, they are often out of pocket for 'indirect costs' of non-PBS medication and equipment, as well as the costs of transport and time away from home or work (Rowbottom et al. 2003: 3; Willis 1995).

Notwithstanding the potential of diabetes to incapacitate populations, accurate information on the number of diabetics in the Pilbara is difficult to acquire. One attempt to estimate this draws on evidence from a host of previous studies to apply a 30 per cent rate to the Indigenous population of the Pilbara aged 25 years and over (Rowbottom et al. 2003: 23). From this an estimate of 813 Indigenous diabetics is derived for the Pilbara in 2001. It is worth noting that Rowbottom et al. considered the 30 per cent rate to be conservative. If we were to update this estimate by applying the same rate to the projected estimate of the Indigenous population in 2006, then the imminent number of Indigenous diabetics in the Pilbara would be 1016. It is worth noting the implications of this finding in terms of potential labour force numbers since this estimated number of diabetics alone in 2006 (to say nothing of other disabling conditions) is almost equivalent to the projected numbers in mainstream employment in the same year (1378).

Also available using hospital statistics are comparative data on Indigenous and non-Indigenous health status between the Pilbara and health service regions in the rest of Western Australia. These data, compiled by the Western Australian Department of Health using hospital separations for the period 1994–2000, detail the comparative rates of the five conditions that account for 75 per cent of all Indigenous deaths in Western Australia as a whole – circulatory disease, cancer, respiratory disease, injury and poisoning, and diabetes (Watson, Ejueyitsi & Codde 2001).

As indicated in Table 7.12, respiratory diseases, and injury and poisoning, account for the majority of hospitalisations among Indigenous people from the East Pilbara, followed by diseases of the circulatory system. However, according to the SRRs, the hospitalisation rate for respiratory diseases was lower than expected compared to the total Pilbara Indigenous population. The same result was observed for diabetes in the East Pilbara. Rates for all other causes were not significantly different compared to the Pilbara total Indigenous population. In terms of comparative rates with the non-Indigenous population of the East Pilbara, the age-standardised rate ratios indicate significantly higher Indigenous rates for circulatory diseases, respiratory diseases, injury and poisoning, and diabetes – in the latter instance, more than eight times higher.

Table 7.12. Indigenous and non-Indigenous hospitalisation statistics for selected major health conditions for East Pilbara, 1990–1999

Condition	Indigenous			Non-Indigenous			
	no.	SRR[a]	ASR[b]	no.	SRR[a]	ASR[b]	ASRR[c]
Circulatory Disease	557	1.1*	43.1	1041	1.1⁻	16.1	2.7 [H]
Cancer	181	1.0*	12.9	1045	1.1*	13.3	1.0 [NS]
Respiratory Disease	2207	0.8^	108.4	2005	1.0*	20.3	5.3 [H]
Injury & Poisoning	2190	1.0*	99.4	2588	1.0*	22.7	4.1 [H]
Diabetes	156	0.4^	10.1	89	1.1*	1.2	8.4 [H]

[a]Standardised hospitalisation rate ratios.

[b]Age-standardised rate per 100 000 population

[c]Ratio between Indigenous and non-Indigenous age-standardised rate.

Key: * = Not significantly different compared to the total Pilbara Indigenous population; ^ = Significantly lower compared to the total Pilbara Indigenous population; ⁻ = Significantly higher compared to the total Pilbara Indigenous population; [H] = Significantly higher than the non-Indigenous population in the region; [NS] = Not significantly different than the non-Indigenous population in the region.

Source: Watson, Ejueyitsi, and Codde 2001.

Among residents of the West Pilbara Health District, the rate of Indigenous hospitalisation for diabetes is significantly higher than the Indigenous rate for the Pilbara as whole, and as much as 65 times higher than the age-standardised non-Indigenous rate (Table 7.13). Hospitalisation caused by respiratory diseases is also much higher in the West Pilbara than in the East, and significantly higher among Indigenous compared to non-Indigenous residents.

Table 7.13. Indigenous and non-Indigenous hospitalisation statistics for selected major health conditions for West Pilbara, 1990–1999

Condition	Indigenous			Non-Indigenous			
	no.	SRR[a]	ASR[b]	no.	SRR[a]	ASR[b]	ASRR[c]
Circulatory Disease	231	0.9*	34.9	920	0.9^	14.6	2.4 [H]
Cancer	94	1.1*	17.8	983	0.9*	12.1	1.5 [H]
Respiratory Disease	2033	1.4⁻	206.3	2360	1.0*	20.5	10.1 [H]
Injury and Poisoning	1371	1.1*	118.4	3003	1.0*	23.3	5.1 [H]
Diabetes	549	2.2⁻	65.1	87	1.0 [#]	1.0	65.1 [H]

[a]Standardised hospitalisation rate ratios.

[b]Age-standardised rate per 100 000 population.

[c]Ratio between Indigenous and non-Indigenous age-standardised rate.

Key: * = Not significantly different compared to the total Pilbara Indigenous population; ^ = Significantly lower compared to the total Pilbara Indigenous population; ⁻ = Significantly higher compared to the total Pilbara Indigenous population; [H] = Significantly higher than the non-Indigenous population in the region; [NS] = Not significantly different than the non-Indigenous population in the region; # Number of cases too low to draw meaningful conclusions.

Source: Watson, Ejueyitsi, and Codde 2001

Risk factors

A proper understanding of the morbidity profile of the Pilbara would examine the complex of underlying related causes, including the risk-taking behaviour of the population (see Interview segment 49, p. 128; Interview segment 50, p. 129; Interview segment 51, p. 129). While this is not provided in full, some insight into the nature of the connections here is provided by the application of aetiological fraction methodology to deaths and hospital data to estimate the incidence of mortality, illness and injuries attributable to alcohol consumption (Unwin et al. 1997). These are shown for the Pilbara SD in Table 7.14 using all deaths for the period 1984–1995, and all separations for the period 1993–1995. With an average of around 10 alcohol-related deaths per year, this represented around 10 per cent of all deaths in the Pilbara over this period. As indicated, these deaths were evenly divided between those due to alcohol-related diseases such as liver cirrhosis, and those due to injuries, especially road injuries. As for hospital separations, the majority of those related to alcohol (68%) presented as injuries due largely to assaults, road injuries and falls. If we examine more recent data on the incidence of alcohol-related morbidity for the Pilbara by Indigenous status (Table 7.15), we can see that Indigenous people accounted for two-thirds of these, and separations among Indigenous females were just as high as for Indigenous males.

Table 7.14. Deaths and hospital separations due to alcohol-related conditions: Total population of the Pilbara, 1984–1995

Alcohol-related conditions	No. of deaths (1984–95)	No. of hospital admissions (1993–95)
Liver cirrhosis	17	44
Alcoholism	15	213
Cancers	4	13
Stroke	8	15
Other related diseases	11	95
Road injuries	35	104
Falls	3	195
Suicide	6	15
Assaults	9	456
Other related injuries	2	49
Total	110	1199

Source: Unwin et al. 1997.

Table 7.15. Alcohol-related morbidity in the Pilbara Population Health Unit area by Indigenous status and sex, 1998–2002.

	Males		Females		Total	
	No.	%	No.	%	No.	%
Non-Indigenous	509	41.2	185	20.9	694	32.8
Indigenous	725	58.8	700	79.1	1425	67.2
Total	1234	100.0	885	100.0	2119	100.0

Source: Pilbara Population Health Unit 2004: Appendix 3.

The importance of these alcohol-related statistics in the context of the present exercise is highlighted by the fact that Indigenous people themselves (at least in Port Hedland) recognise a strong connection between substance misuse and economic marginalisation (Saggers & Gray 2001: 66), and this is certainly a strong message to emerge from the interviews held with respondents from across the Pilbara as part of the present study (see Interview segment 1, p. 57; Interview segment 11, p. 60).

Disability

One element of health status that can have a direct impact on the capacity of individuals to participate in economic activity is disability, defined as any continuing condition that restricts everyday activities. Such restriction can be due to an intellectual, cognitive, neurological, sensory or physical impairment or a combination of these; it may also be permanent or episodic in nature. However, with appropriate aids and services the restrictions experienced by many people with a disability may be overcome. Overall, in Western Australia, the most recent measure of the labour force participation rate of adults with a disability indicates that this is surprisingly high at 56 per cent, although this compares to 80 per cent among those without a disability (ABS 1998).

Establishing the number of people in the Pilbara with a disability using public access information is difficult, especially if the aim is to do this by Indigenous status. In 1998, the Western Australia Disability Services Commission (DSC) determined the overall numbers in the Pilbara with a disability. In all, a total of 5305 people (12% of the Pilbara population) were found to have an activity restriction. One-third of these had a profound or severe disability, though the majority (40%) were moderate/mild (DSC 1998). Most of those with disabilities (67%) experienced a physical restriction on their activities.

One way to estimate the extent of disability among Indigenous people in the Pilbara is to refer back to Table 4.8 which indicates that 510 disability payments in 2005 were to Centrelink customers who identified as Indigenous. As a proportion this would represent only 9 per cent of all disabled persons in the Pilbara using the 1998 DSC figures as a guide. As such, this is likely to be a substantial underestimate on the basis of population share and relative health status. Given a current Indigenous population share of 18 per cent, we may confidently double the estimate to at least 1020. Just what the labour force status of this group might be remains unknown. However, if the labour force status data from the 1998 ABS Disability Survey for the whole of Western Australia were applied, then almost half of these (448, or 44%) would be included with those not in the labour force as indicated in Figs. 3.7 and 3.8.

Indigenous perspectives

Interview segment 44

Ten or twenty years ago I used to be into alcohol and being in trouble with police. Whilst in prison I decided, 'what am I doing in here, my family all outside, and my kids?' That was a turning point. I wanted to get my people back to the country, to where I am now on my block. But to do that I had to survive, I had to look for work to keep my family going. I had to look for a job up here. I had a job at a medical centre taking people out for hunting and give them a feed. But there was no place for them to go back to, and they knew they would end up doing the same. What we wanted was an outback alcohol rehabilitation place, which ended up being built. It is still there today but its name changed from being an alcohol rehabilitation community to just a community, and we moved the program back into town. It was better out bush.

Interview segment 45

Health is pretty good here. I go and see the hospital. They used to come around and take old people's blood pressure. There are a lot of old people in the town that need medical attention like medication, you know once a week, what's once a week to go around and check old people's blood pressure, sugar level and see if they got all their medication and pill boxes up to date? They used to come around like that. One of them nurses was really good, she used to come up in the street and say hello and ask how I am. She used to talk to young people too about how they feeling.

Interview segment 46

But don't talk to me about dentist! You got to go to Newman, its free one, or Karratha. They got a dentist coming here to Tom Price from Perth but it costs an arm and a leg, even just for a five-minute visit, even just to peek in your mouth! People just don't go because they can't afford it. When it gets to be a real problem we just go to hospital for dental things. And they send you to Hedland or Newman. My little grand-daughter here, she had a teeth trouble. They told us you gotta' go Hedland or you gotta' go Mt Newman. Sometimes you can get travel assistance if you fill in that form for travel assistance, but sometime it can happen in an off week and people got no money to travel, or for medicine.

Interview segment 47

Education and health doesn't just affect Aboriginal people, it affects everyone in this day and time. Yeah we have a nurse who comes out to our community every week and a doctor who comes out every fortnight and we also got our own Aboriginal health worker and Gumala are setting up an office for her, that's all starting to happen.

Interview segment 48

We don't have so much of a problem with health – common colds, runny nose, eyesores, and ear problems, just childhood things that affect everyone. We have good health delivery service. The health workers put their heart into it and they get to know the communities, and are really good. We not only have faith and trust and respect in them, but we look forward to when they are going to come again. The health workers are the best. But sometimes we just get to know them and then they are gone somewhere else, and we lose that contact.

But the problems in health are amongst those above them, the policy and management mob who don't want to go down to the grass roots level, they just send the health workers out, and don't visit themselves. I call them the shiny-bums. And the health system is not held accountable, There is so much energy put into it from the funding body and so much energy put into it from the service delivery, but when it gets down to the results it's not achieving those outcomes. It's too much massage, and healthcare is only during 8am and 5pm, but it's all the time for us.

Interview segment 49

A lot of that 20 to 30 year age group think that it is easier to spend their money on grog and live off somebody else. They rely on parents and grandparents and just move house to house. I've been observing this for many years. I don't think that Roebourne is a negative town, I just know where the negative behaviors are. Its the same in every other town. Its about the nature of alcoholism, not the nature of Aboriginal people. When you go back in history, our people used to walk around in the bush, they were 24 hours a day out there, now we are 24 hours a day in town in an urban setting, fast food and processed food, dependency on money to be able to get the food rather than hunting and gathering. Health standard is down, and there is a whole range of reasons. There's not enough clear educational awareness about health and what people mean when they talk about health. Someone might say, 'well I don't go to hospital so therefore I am healthy', but that may not be true. People don't recognise that all the things they do like diet may be impacting them badly and that even if they haven't been yet, they may well be on a one way trip to the hospital and never coming back! There's a lot of knowledge around, but not an educational approach to it. You can go to Centrelink and fill out forms and learn how to get finance. There's no place where you can go and sit down and get advice on your health before it happens. You can go to Community Health but they aren't really doing that either. With the sort of health issues my people have, it's no good to sit back and wait for people to come to you when they are sick, they should be reaching out to the people and making themselves more available. They should come out and visit. A visit means 'I care', sitting in your office means 'you come and see me if you got a problem, otherwise I'm not really worried about you'.

Interview segment 50

My daughter looks after me and takes me hospital and also Wirraka Maya. I also gotta son live with me. He been come back from hospital in Perth yesterday, he's really sick. He got low sugar, high blood pressure, and he got to take medicine for his head all the time. He was a sniffer, that petrol destroyed a lot of our young people. He's just a young man, but he got old body, can't talk and can't look after himself, he's a walking time bomb. I got to watch watch watch all the time, in case he wanders off, or gets into grog or something. Sometimes he goes crazy and we got to call police. I get little bit of money for him from the trustee, buy him smoke and food, but I got to ask for more if he needs clothes or shoes. He got to go clinic all the time, and I got to take him. We can't walk there, too far, we got no motorcar, so we take taxi. I got that low sugar too, and I can't hardly walk! We living with relative now, because we got kicked out of Homeswest house, too much damage, and now we got to pay $2000. That health mob come around too, they really good.

Interview segment 51

In the community all the problems come from that like domestic violence, this one up here, he's into ganja too. When they are sober, not drugged up, they get agitated. Cigarette smokers are like that too. Getting really tense, makes their imagination go wild, and a lot of them we are starting to see are going into schizophrenia, that's that paranoia, then they get sent away and they get needles. There is drug and alcohol education. I think the youth center is starting to do it, and they do counselling. There is one just for this Ashburton area and she's based in Newman. People here are free to do it, but there are not enough resources in that area at the moment. We could get that program here, but the people gotta be willing to. The only way we can really force them is to start to bring in laws in the community to make people participate in programs. We trying to make this a good community and the only way we can start to force them, cause while the mob sitting down even those people sitting down on the dole and stuff in the community, they should be getting involved.

8. Crime and justice

Interaction with the police, and subsequently with the courts and various custodial institutions, is a pervasive component of Indigenous social and economic life in the Pilbara region. In the 1994 National Aboriginal and Torres Strait Islander Survey (NATSIS), an estimated 18 per cent of Indigenous people aged 13 years and over in the Ngarda Ngarli Yarndu ATSIC Region reported that they had been arrested by police in the previous five years (ABS 1996a: 70). This amounted to some 570 individuals. The equivalent proportion in the Warburton ATSIC Region (the northern half of which falls within the Pilbara SD) was 23 per cent representing some 420 individuals (ABS 1996b: 70). In Western Australia, as a whole, the rate was 25 per cent. Interestingly, in the same survey, 72 per cent of respondents in the Ngarda Ngarli Yarndu ATSIC Region perceived family violence to be a problem in the local area. This was the fourth highest rate of any ATSIC Region in the country and way above the rate in Warburton (35%). In both of these Pilbara ATSIC Regions, the primary reasons given for arrest included (in descending order) disorderly conduct, drink driving offences, assault, and outstanding warrants. In the decade since this survey, little seems to have changed (see Interview segment 57, p. 143; Interview segment 58, p. 143).

A note on data sources

Crime statistics in Western Australia are available from a variety of sources reflecting different stages of interaction with the criminal justice system. The initiating factor, of course, is contact with the police either by way of crime reporting, or via an apprehension (arrest), or a summons. Thus, the profile presented here does not represent all criminal activity, only that processed by the criminal justice system. Nonetheless, such processing yields a range of data concerning the nature of offences and offenders with separate reporting for juveniles (aged 10–17 years), and adults (aged 18 and over). Individuals who are charged with an offence are further processed by the courts (a charge being an allegation laid by the police before the court or other prosecuting agency that a person has committed a criminal offence). Statistics relating to the activities of the lower courts are captured by the Department of Justice CHIPS (Childrens Court and Petty Sessions) database. As for those charged who are found guilty of an offence, imprisonment data are available from the Department of Justice's Total Offender Management System (TOMS), while non-custodial community corrections data can be extracted from the records of the Community and Juvenile Justice division of the Department of Justice. Data regarding those held in police lock-ups are provided via the Western Australia Police Lock-up Admissions System which records all admissions to and exits from police lock-ups across the state.

The Crime Research Centre (CRC) at the University of Western Australia has access to all of these data for analysis and reporting under agreements with the Western Australia Police and Western Australia Department of Justice. Using this access, the CRC produces an annual comprehensive compendium of crime and justice statistics for the state – *Crime and Justice Statistics for Western Australia* – detailing the nature and pattern of offences and sentences, and the characteristics of offenders and those sentenced. Among the characteristics explored is ethnicity, and the basic ethnic classification employed by the CRC in its reporting is Indigenous/non-Indigenous. However, the manner in which Indigenous status is determined varies between police and courts data. In the Police Offence Information System (P49), 'ethnic appearance' is a term used to describe the visual appearance of victims and offenders. The field is completed on the basis of the attending police officer's subjective assessment of the person's appearance, and is recorded for operational purposes only. As the CRC cautions, given the subjective nature of the assessment upon which these data are based, it is possible that a person attributed to a particular group does not belong to that group. Data from the lower courts presents far greater difficulty in terms of establishing Indigenous participation in the criminal justice system since in Western Australia as a whole the Indigenous status of defendants is unknown in 85 per cent of cases (Loh & Ferrante 2001: 20).

Reported crime

The most common crimes reported to police in Western Australia are classified as offences against property (including burglary, property damage/arson, and motor vehicle theft), and offences against the person (including assault, sex offences, and robbery). Other less commonly reported crimes include drug offences, fraud and receiving, and good order (mostly trespass and vagrancy), while other sundry offences (mostly offences against justice procedures) make up the remainder (Fernandez & Loh (2001: 10–12). Table 8.1 shows the relative distribution of offence rates by type of offence for the total population of each of the Pilbara Shires in 2003. Clearly, the highest overall crime rates occur in Port Hedland Shire, and this is so for the individual categories of property offences, and other offences such as justice procedures and good order. East Pilbara records relatively high rates of offences against the person, especially assault, while Roebourne stands out for property offences. The lowest rates, for all types of offence are recorded in Ashburton (see Interview segment 51, p. 129;Interview segment 52, p. 142; Interview segment 53, p. 142; Interview segment 55, p. 142; Interview segment 56, p. 143).

Table 8.1. Reported offence rates[a] by type of offence: Pilbara Shires, 2003

	Port Hedland	Roebourne	Ashburton	East Pilbara	Total Pilbara
Against the person					
Homicide	0.3	0.0	0.2	0.0	0.1
Assault	30.0	15.1	6.7	26.0	19.5
Sexual offence	4.5	3.3	1.2	4.3	3.0
Robbery	0.9	0.4	0.2	0.0	0.3
Other	0.5	0.2	0.2	6.1	2.6
Sub-total	36.1	19.0	8.5	36.4	25.5
Property					
Burglary (res)	40.7	16.6	13.3	9.4	18.1
Burglary (non-res)	13.0	14.2	11.4	8.0	13.0
Vehicle theft	11.3	3.1	2.8	3.8	7.1
Theft from vehicle	23.2	16.1	8.0	6.4	14.5
Other theft	83.1	67.2	41.0	46.1	58.9
Damage	44.3	31.5	25.3	28.9	30.0
Sub-total	215.5	148.7	101.7	102.5	141.5
Drug offences					
Sub-total	7.4	8.3	5.4	10.2	9.4
Other offences					
Justice procedures	5.1	2.4	0.5	2.6	3.0
Good order	10.0	8.4	2.6	3.6	5.4
Sub-total	15.1	10.8	3.5	6.2	8.4
Total	274.1	186.8	119.0	155.4	184.9

[a]Rate per 1000 persons.

Source: Western Australia Office of Crime Prevention, Community Safety and Crime Prevention Profile.

If we break these offences down into more detailed characteristics, the spatial pattern becomes more varied (although Port Hedland still stands out with the highest rates), and a clear difference in rates emerges between Indigenous and non-Indigenous residents of the Pilbara. Table 8.2 shows the rate of property offences for different urban locations across the Pilbara SD according to the Indigenous status of the victim. A consistent pattern across the Pilbara is that non-Indigenous people are far more likely to report property offences than Indigenous people. This is especially so in Karratha and Wickham where the rate of property offences against non-Indigenous people is twice that against Indigenous people.

Table 8.2. Property offence rates[a] by Indigenous status: Pilbara SD, 2001

	Indigenous	Non-Indigenous	Ratio
Karratha	33.6	60.5	0.56
Marble Bar	11.1	15.0	0.74
Newman	55.1	65.8	0.84
Pannawonica	0.0	5.3	0.00
Paraburdoo	11.6	16.8	0.69
Port Hedland	83.3	120.2	0.69
Roebourne	34.3	129.4	0.27
Tom Price	50.7	55.7	0.91
Wickham	25.6	64.6	0.40

[a]Rate per 1000 persons.
Source: Fernandez 2003.

By contrast, the more likely victims of violent offences are Indigenous people (Table 8.3). In Newman, the rate at which Indigenous people report violent offences is four times the non-Indigenous rate, while in Port Hedland it is three times as high. In the latter case, the actual rate of violent offence reporting by Indigenous people is by far the highest in the Pilbara, involving as much as 10 per cent of the Port Hedland Indigenous population.

Table 8.3. Violent offence rates[a] by Indigenous status: Pilbara SD, 2001

	Indigenous	Non-Indigenous	Ratio
Karratha	18.5	14	1.32
Marble Bar	17.5	8.5	2.06
Newman	83.7	20.2	4.14
Pannawonica	0.0	2.4	0.00
Paraburdoo	11.6	5.5	2.11
Port Hedland	104.4	35.8	2.92
Roebourne	52.0	58.1	0.90
Tom Price	20.3	9.2	2.21
Wickham	41.0	18.2	2.25

[a]Rate per 1000 persons.
Source: Fernandez 2003.

When it comes to arrest rates, the dominance of Indigenous people within the regional criminal justice system becomes overwhelmingly apparent. Overall in the Pilbara, in 2001, a total of 898 distinct Indigenous people were arrested (Loh & Ferrante 2001: 16). This represented 14 per cent of those aged 10–18 years, and 23 per cent of those aged 19 and over. Within the Pilbara, these rates varied considerably according to offence type with the highest Indigenous arrest rates for property offences reported in Port Hedland, although compared to non-Indigenous rates the greatest relative difference was reported from Wickham where the Indigenous arrest rate was 74 times greater, albeit derived from a small base (Table 8.4). The lowest rates were reported in Pannawonica and Paraburdoo.

Table 8.4. Arrest rates[a] for property offences by Indigenous status: Pilbara SD, 2001

	Indigenous	Non-Indigenous	Ratio
Karratha	53.8	4.3	12.5
Marble Bar	28.6	0.0	n/a
Newman	106.1	2.7	39.3
Pannawonica	0.0	0.0	0.0
Paraburdoo	7.7	5.2	1.5
Port Hedland	128.5	6.8	18.9
Roebourne	81.6	18.8	4.3
Tom Price	87.8	2.7	32.5
Wickham	82.0	1.1	74.5

[a]Rate per 1000 persons.
Source: Fernandez 2003.

If we focus on relative arrest rates for just two other types of offence, we can see that Indigenous people are arrested for violent offences at a consistently higher rate than non-Indigenous people in all parts of the Pilbara (Table 8.5), especially in Newman and Tom Price, while arrest for good order offences is mostly focused on Newman, Marble Bar and Port Hedland (Table 8.6), with the latter recording very high rates for Indigenous people (34% of the population).

Table 8.5. Arrest rates[a] for violent offences by Indigenous status: Pilbara SD, 2001

	Indigenous	Non-Indigenous	Ratio
Karratha	15.1	3.3	4.6
Marble Bar	17.5	0.0	n/a
Newman	67.4	2.2	30.6
Pannawonica	9.3	1.8	5.2
Paraburdoo	7.7	3.2	2.4
Port Hedland	70.3	4.3	16.3
Roebourne	44.9	18.8	2.4
Tom Price	50.7	2.0	25.4
Wickham	25.6	2.2	11.6

[a]Rate per 1000 persons.
Source: Fernandez 2003.

Table 8.6. Arrest rates[a] for good order offences by Indigenous status: Pilbara SD, 2001

	Indigenous	Non-Indigenous	Ratio
Karratha	42.0	7.6	5.5
Marble Bar	58.7	0.0	n/a
Newman	193.9	5.2	37.3
Pannawonica	18.5	4.4	4.2
Paraburdoo	3.9	1.9	2.1
Port Hedland	343.9	9.2	37.4
Roebourne	41.4	33.4	1.2
Tom Price	40.5	3.7	10.9
Wickham	25.6	11.4	2.2

[a]Rate per 1000 persons.
Source: Fernandez 2003.

Contact with the police

Contact between the police and the regional population is recorded as persons are apprehended by the police (either via arrest or summons), or are diverted (as juveniles) through the cautioning system and referred to juvenile justice teams. Apprehensions data are derived from the police P18 form and describe offences charged by police either via arrest or summons. According to data reported by the CRC, Indigenous people accounted for 73 per cent of all apprehensions by police in the Pilbara in 2003 (Fernandez et al. 2003: 49). In terms of distinct Indigenous persons, this amounted to a total of 979 arrests, with 750 of these among adults (aged 20 years and over) and 229 among youth (aged 10 to 19 years). These figures represented 23.6 per cent of all Indigenous adults and 18.6 per cent of Indigenous youth (Fernandez et al. 2003: 50).

The Office of Crime Prevention (OCP) also provides data on the number of unique offenders arrested. In the Pilbara as a whole, a total of 1740 unique persons were arrested in 2003 and the OCP reports that fully 60 per cent (1047) were Indigenous, with males accounting for 78 per cent of these (Table 8.7). To put this in a regional economic context, the total number of Indigenous people arrested is almost equivalent to the total number of Indigenous people aged 15–54 estimated to be employed in the regional mainstream labour market in 2006 (1378). Indeed, if we narrow the focus to consider particular cohorts, the full potential economic impact becomes even more apparent. Thus, a total of 482 Indigenous male individuals aged 18–34 years were arrested at least once during 2003. This comprised as much as 49 per cent of the 2001 ERP in that age group which is exactly the same proportion as those employed in the 15–34 age group. Indeed, cross-reference to the relatively poor labour force status of Indigenous people between the ages of 15 and 34 (see chapter 3), suggests the likelihood that high arrest rates are a major barrier to regional participation. This is not surprising given the disruption to labour market engagement that contact with the police and its subsequent consequences are likely to imply.

Table 8.7. Unique offenders arrested by sex and Indigenous status: Pilbara SD, 2003

	Males	Females	Total
Indigenous	813 (58.6)[a]	234 (66.5)	1047 (60.1)
Non-Indigenous	575 (41.4)	118 (33.5)	693 (39.9)
Total	1388 (100.0)	352 (100.0)	1740 (100.0)

[a]Percentages in parentheses.

Source: Western Australia Office of Crime Prevention, Community Safety and Crime Prevention Profile.

Examined by sub-region within the Pilbara (Table 8.8), we can see that the number of individual offenders varies quite markedly between the different Shires, especially among males, with Port Hedland Shire displaying the largest number of individuals arrested and by far the highest male rate with as much as 43 per cent of Indigenous males over age 10 arrested at least once in 2003.

Table 8.8. Unique Indigenous male and female offenders arrested as a proportion of population aged 10 and over: Pilbara Shires, 2003

	Males		Females		Total	
	No.	% of 10 + population	No.	% of 10 + population	No.	% of 10 + population
Port Hedland	353	42.8	108	11.6	461	26.3
Roebourne	189	23.9	40	5.9	229	15.6
Ashburton	69	23.1	20	7.4	89	15.7
East Pilbara	202	32.6	66	10.4	268	21.4
Total	813	32.1	234	9.3	1047	20.8

Source: Western Australia Office of Crime Prevention, Community Safety and Crime Prevention Profile.

Data on admissions to police lock-ups in different parts of the Pilbara also provide for a more detailed examination of the geographic spread of contact with the police. Reasons for admission to lock-ups include arrest (apprehended and charged by police but not sentenced), drunken detainee, fine default, remand, sentenced, and held under a warrant. Receptions for public drunkenness have typically accounted for a large component of all lock-ups for Indigenous people, although this has declined steadily in Western Australia as a whole since 1996 (Loh & Ferrante 2001: 35–6). At face value, the data shown in Table 8.9 might suggest some link between alcohol misuse and admissions to lock-ups. This is revealed by the very low number of admissions indicated for Port Hedland where Saggers and Gray (2001: 49) find a strong association between the growing use by Indigenous people of services provided by sobering-up shelters and the numbers admitted to police lock-ups. Having said that, they also note the same relationship in Roebourne (Saggers & Gray 2001: 45), but this does not seem to be reflected in Table 8.9. Equally, the low numbers at Port Hedland lock-up are more than compensated for by the very large intake at the South Hedland lock-up. Clearly, South Hedland stands out as having the highest number of

Indigenous admissions to lock-ups, while Newman displays the highest ratio, with Indigenous admissions there almost seven times higher than for non-Indigenous people.

Table 8.9. Admissions to police lock-ups in the Pilbara SD, 2001

	Indigenous	Non-Indigenous	Ratio
Karratha	172	218	0.8
Marble Bar	73	1	73.0
Newman	341	50	6.8
Paraburdoo	3	15	0.2
Port Hedland	6	15	0.4
Roebourne	189	71	2.7
South Hedland[a]	1430	351	4.1
Tom Price	28	10	2.8
Total	2243	731	3.1

[a]South Hedland figures supplied separately by John Fernandez, Crime Research Centre, University of Western Australia.
Source: Fernandez 2003.

Lower Courts data

Data were obtained from the Western Australia Department of Justice regarding the number of adjudicated cases from the Children's Courts and Courts of Petty Sessions in the Pilbara for the years 2000 to 2003 inclusive. A major drawback for the analysis of these data for the present exercise is the lack of a process in court reporting of ethnic self-identification. As a result Indigenous status is unrecorded for 22 per cent of distinct persons appearing before the lower courts, and for 13 per cent of those before the Children's Court in 2003 (Fernandez et al. 2003: 75, 107). However, given the arrest rates described above, it can be reasonably assumed that the majority of these unrecorded cases refer to Indigenous persons. Net of these 'not stateds', Indigenous people accounted for 23 per cent of all individuals appearing before the Courts of Petty Sessions in Western Australia in 2003, and 40 per cent of all juveniles before the Children's Court.

Table 8.10 shows the distribution of offences heard by the children's courts and courts of Petty Sessions in the Pilbara between 2000 and 2003 according to the type of offence. By far the largest number of offences (27%) in the Children's Court were for unlawful entry/burglary and break and enter, followed by offences against justice procedures. In the Lower Courts, road traffic and motor vehicle regulatory offences predominate followed by public order offences and offences against justice procedures (see Interview segment 54, p. 142; Interview segment 55, p. 142). If just these three categories were eliminated, then the number of offences heard in the Pilbara by the Courts of Petty Sessions would be reduced by 55 per cent. Of course, road traffic and motor vehicle offences are of particular significance as they have the potential, in the form of suspended licenses or outstanding fines, to exclude many Indigenous people from employment in mining and other sectors that involve driving.

Table 8.10. Annual average distribution of offences heard in the Children's Court and Court of Petty Sessions by offence type: Pilbara SD, 2000–2003

ASCO Division	Children's court		Petty Sessions	
	No.	%	No.	%
Abduction and related offences	1	0.2	1	0.0
Acts intended to cause injury	30	9.0	241	9.7
Dangerous/negligent acts	25	7.7	396	15.9
Deception and related offences	1	0.2	29	1.2
Homicide and related offences	1	0.2	2	0.1
Illicit drug offences	9	2.8	105	4.2
Miscellaneous offences	2	0.5	51	2.1
Offences against justice procedures	52	15.7	441	17.8
Property damage	17	5.2	81	3.2
Public order offences	33	10.0	421	17.0
Road traffic/motor vehicle regulatory offences	22	6.5	507	20.4
Robbery, extortion and related offences	2	0.5	1	0.0
Sexual assault and related offences	2	0.6	18	0.7
Theft and related offences	44	13.2	112	4.5
Unlawful entry/burglary, break and enter	89	27.1	45	1.8
Weapons and explosives offences	2	0.5	33	1.3
Grand Total	329	100	2481	100

Source: Western Australia Department of Justice.

As for the findings of court proceedings in the form of penalties (sentences), these can be grouped into four broad categories: custodial, non-custodial, fines and dismissals. According to the ABS sentence type classification (ABS 2003: 71), custodial orders involve custody in a correctional institution as life imprisonment, imprisonment with a determined term, or periodic detention. They also include custody in the community under an Intensive Corrections Order or home detention. Suspended sentences also fall under custodial orders. Non-custodial orders include a variety of community supervision or work orders and community service orders, as well as probation and treatment orders. Other non-custodial orders include good behaviour bonds and recognisance orders, while monetary orders basically refer to fines or recompense to victims as well as licence disqualification/suspension/amendment and forfeiture of property.

As non-custodial sentences are the most common it is worth defining some further aspects of these. For example, Community Based Orders (CBOs) allow the court to order an offender to be managed by a Community Corrections Officer for the purposes of any one or more requirements of supervision, community service of between 40 to 120 hours, and/or programs aimed at the offender's behaviour. Intensive Supervision Orders (ISOs) are similar but provide for longer and more stringent supervision including curfews. Work and Development Orders (WDOs) are the last option prior to imprisonment for people who are in default of a fine. The order requires that the offender perform a specified number of hours of community work and personal development.

In the Pilbara, as in all remote Indigenous communities in Western Australian, these non-custodial orders are carried out under the Indigenous Community Supervision Agreement which offers communities a key role in the decision making about offender management. As Parriman and Daley (1999) point out, communities decide themselves whether to accept an offender under supervision, they determine the most appropriate person to administer the supervision order, and they are largely responsible for determining the supervision regime. One consequence has been a tendency on the part of the courts to make greater use of community-based sentencing (Parriman & Daley 1999: 3), and this is reflected in the sentencing data.

Table 8.11 shows the distribution of penalties awarded to convicted charges between 2000 and 2004 by the Children's Courts in the Pilbara according to the type of penalty awarded. By far the largest number of convictions in the Children's Courts (57%) attracted a community-based order, with a monetary fine in almost one-fifth of awarded cases. The actual number of custodial orders served is relatively small. This is not the case in the lower courts as shown in Table 8.12. Almost 13 per cent of cases (350) attracted a custodial sentence or suspended imprisonment order, with community based orders far less prevalent. However, monetary fines were by far the largest single penalty awarded.

Table 8.11. Annual average distribution of court penalties awarded in Children's Courts: Pilbara SD, 2000–2004

Outcome	No.	%
Custody	12	4.6
Suspended Imprisonment Order	1	0.4
Community Based Order	148	56.7
Fine	49	18.8
Good Behaviour Bond/Recognisance	21	8.0
No Punishment	25	9.6
Adjourned	5	1.9
Total	261	100.0

Source: Western Australia Department of Justice customised table.

Table 8.12. Annual average distribution of court penalties awarded in Courts of Petty Sessions: Pilbara SD, 2000–2004

Outcome	No.	%
Custody	212	7.7
Suspended Imprisonment Order	138	5.0
Community Based Order	434	15.8
Fine	1719	62.5
Good Behaviour Bond/Recognisance	141	5.1
No Punishment	31	1.1
Adjourned	75	2.7
Total	2750	100.0

Source: Western Australia Department of Justice customised table.

Working out the impact of these court penalties on the Indigenous population of the Pilbara is no easy task given the lack of self-identified Indigenous status in court reporting. However, between July 2001 and June 2002 a total of 286 distinct Indigenous persons (84% of them males) who were received into Western Australian prisons indicated that the Pilbara was their usual place of residence (Western Australia Department of Justice 2002: 8). Over the same year, a total of 571 distinct Indigenous persons from the Pilbara (73% of them males) were also served with Community Corrections Supervision orders by the lower courts. At any one time, however, the numbers actually in detention or serving such orders was much lower than this. For example, the Department of Justice Census of Prisoners indicated that a total of 88 Indigenous prisoners who were in custody on the night of 30 June 2002 had a prior usual address in the Pilbara, while this also applied to 95 Indigenous individuals on supervision orders (Western Australia Department of Justice 2002: 20, 44).

This prison figure is interesting as it is considerably lower than the total of 157 Indigenous persons with a usual residence address in the Pilbara that was recorded as enumerated in a prison (anywhere in Australia) at the 2001 Census. This figure amounted to 4.6 per cent of the census population aged 18 years and over. If we apply this same proportion to the projection of the population 18 years and older for 2006 we can produce an estimate of some 200 Indigenous persons currently in prison at any one time. If we then add to this an estimate of 113 Indigenous adults in 2006 on community supervision orders at any one time (using as a guide the 2002 figure from the Department of Justice reported above) then overall around 313 adults may currently be detained in some form. Of course, this is simply the stock, whereas as the flow through custodial/non-custodial sentencing over a given year would be much higher. Clearly, the impact on simple availability to participate in the regional workforce is substantially hampered by this enforced withdrawal of labour, to say nothing of the lingering negative effects of incarceration. There may, of course be positive impacts of rehabilitation, but these are unquantified here. Once again, it is useful to place these estimates against the likely breakdown of the working-age population by labour force status as shown in Figs. 3.7 and 3.8. On this reckoning, almost one-fifth (18%) of those not in the labour force are under custodial and non-custodial sentences, a proportion that is likely to be much higher among males.

Indigenous perspectives

Interview segment 52

Alcohol abuse is the main issue towards people going to jail, and drugs, and that will push them towards stealing because they want another fix. People around here don't leave their keys in the car because otherwise they'll have no car, and they break into your home, they start doing it in broad daylight now. Stealing from people while they are out working.

Interview segment 53

Well there's been some trouble here with that young couple over there. He bashed her and now the police are here. They staying with her grandmother and she just can't take it and they are damaging her house and when she kicks them out they go and stay with the other nana's house, and now the nana's fightin' too, one sticking up for the woman and one sticking up for the man. A lot of young fellas bash their women to look tough, that's what it's all about, image. And he's been gettin' away because he hasn't got family here and all the men they haven't been sortin' him out and they should be, but they all growl at him and stuff but he's just not listening to anyone. And it's her fault as well, she has to press charges, and she hasn't done that, so might be this time she will but we don't know. So if the woman gonna keep going back to her husband then we can't help her out. It's also jealousy, you know jealousy, run in families here. Man want thing he'll be watchin' the woman's eyes where she looking and that, you gotta be careful of that stuff.

Interview segment 54

Lot of local mob locked up in that Roebourne jail. There aren't enough visits to communities by the cops. You know when there are changes to road rules and things like that? They never send anyone out, or any information out. Lot of people end up in trouble for speed, and vehicle related offences.

Interview segment 55

Lots of young people get into trouble with the police. Most of them going for dd [drink driving], no licence, unregistered motor vehicles, and then not turning up for court, usually they are somewhere else.

Interview segment 56

Well I tell ya, might be drink drink drink, and they go with somebody else and somebody else might be telling them do this one do that one, stole a motor car or something, and then they get in trouble. Take this young fella now. He been assault that same woman. My grandson, police came around this morning and picked him up, and I asked what's he done and they said 'oh, just some problem'. But I want to know because I'm his guardian. I told him you know, 'I want to know what's going on I'm his guardian, bad or good'. 'Oh we just take him to police station and talk with him there.' They didn't tell me why. That's the feelings that hurt me, you don't know what the police going to do down there, I'm thinking all them things, you gotta think about all them things and I tell 'im I might as well ring his father, they said, 'no he's adult now', and I said, 'no way, Aboriginal got a different feelings.'

Interview segment 57

There are multiple reasons for people to end up in the criminal justice system. We've got a society who has gone from being a really good working force, in the past I'm talking, to a society of people who have gone into dependency in the welfare system. And on top of that, you got the substance abuse, you know alcohol and drugs. The loss of discipline, the loss of identity, basically in terms of their background, like law and culture. Loss of identity is not caused by an offending behavior. That identity loss is prior to all that I believe, and part of the cause of offending behavior. Law and culture were things that were very close to the heart of the old people who were probably mostly illiterate, but they were the hard workers, if you compare them to our society today. We have populations of Indigenous people who have grown up with welfare support, which hasn't then produced a workforce attitude or mentality. With that goes alcohol and drugs, and gambling and stuff like that, and during that period the parenting is not happening with the kids. So there is no boundary setting, discipline or skills being passed on to the young people from the parent home. It's a broken parent home.

Interview segment 58

We don't have a program that is suitable for keeping juveniles out of detention, like they have in some other places. We have JJ (juvenile justice) officers who work with juveniles on a one-to-one basis. They don't carry out a specific program as such with them, they either put them into counselling, or place them in a centre to do community work, or centres where they can get involved in an art group or something like that. There is no specific program they can be put through. You have to offend before you can access these services. I haven't seen a program that actually works for people, but there are the requirements of the court and the department. It's not fair to say that they aren't coming up with ideas too. But I know where the problem lies, the problem is here, in the person's heart. Those people who sit in the prison system have been through so many programs that they could probably run an alcohol rehabilitation workshop better than people like us! But when they come out they don't apply it through the processes of the heart. It comes down to personal responsibility.

9. Implications for regional development

The analysis in the preceding chapters details the relative social and economic status of the Pilbara Indigenous population at the commencement of major expansion in the mineral resources sector and associated regional impacts. In the immediate context, it provides an essential quantum to discussions of need, aspirations, and regional development capacities for Indigenous, corporate, and government stakeholders. In future contexts, it provides a benchmark against which the success or otherwise of intended and unforeseen impacts may be measured. Inevitably, and purposely, it constitutes a cross-sectional representation of conditions at the beginning of the twenty-first century, although, where possible, comparison is drawn with the prior situation of Indigenous people at the outset of the contemporary period of Pilbara mining development in the 1960s.

The basic message conveyed is that little has been achieved over the past four decades in terms of enhancing Indigenous socioeconomic status. Progress is now possible on the basis of planned economic development and corporate interest in pursuing Indigenous engagement, but major efforts are required from all three broad stakeholder groups (Indigenous organisations, miners and governments at all levels) in order to ensure that this occurs. The primary dynamic dictating this imperative is the fact of sustained Indigenous population growth against a background of low Indigenous economic status and limited human capital for mainstream economic participation.

Demography

It goes without saying that Indigenous people have by far the longest and most enduring presence in the Pilbara. It is equally true that, aside from the initial upheavals and demographic impacts of sustained contact with Europeans that commenced in the late nineteenth century, the period since the 1960s has seen major shifts in the demographic make-up of the region. For one thing, viewing the Pilbara as a whole, Indigenous people have now become a minority in their own lands following the influx of a predominantly non-Indigenous industrial workforce. For another, significant distributional change has occurred insofar as a formerly widespread Indigenous population has become relatively concentrated in coastal towns and urban centres, though with some recent return to traditional lands. However, the key issue at present, and increasingly into the future, is that regardless of what transpires in terms of regional economic fortunes, the Indigenous population of the Pilbara is set to expand for decades to come. Numerically, the focus will be in growth at younger ages; proportionally, it will occur mostly at older ages. In combination, these expanding cohorts present major challenges for social and economic policy.

As for the non-Indigenous population, major questions surround future (and even current) numbers. The primary variable here is labour demand as dictated by mining and related economic developments, together with the composition of associated workforces in terms of construction-phase, FIFO, and resident components. While the indications are for a renewed increase in non-Indigenous numbers (after recent decline), current demographic parameters indicate that despite this the Indigenous share of the regional population will increase over the next decade to a point approaching one-fifth of the total.

Jobs and economic status

Despite 40 years of substantial economic development in the Pilbara region, the labour force status of Indigenous Pilbara residents has barely altered. While the numbers in work have undoubtedly increased, so has the size of the working-age population. As a proportion then, the Indigenous employment rate remains well below 50 per cent. This contrasts sharply with the rest of the Pilbara population whose primary reason for being in the region is to work. It is even more stark when set against the FIFO and temporary components of the regional workforce whose social and economic allegiances lie elsewhere – a situation not lost on many local Indigenous people as noted in a number of the interviews.

Also apparent from many of the interviews, and confirmed to some degree by the age distribution of labour force status, is a sense that the past 40 years have witnessed a generational attrition in terms of economic engagement as the trades skill-base is relatively focused on older adults, while many younger people find themselves ill-equipped for workforce participation due to low literacy and numeracy, lack of qualifications and work experience, substance misuse and, consequently, low motivation (see also George 2003).

A number of crucial questions arise out of all this for both the mining industry and Indigenous stakeholders. First of all, is the local supply of Indigenous labour sufficient to meet the employment targets that have been set? Second, what is the composition of potential labour supply in terms of human capital and related work-readiness? Finally, what is the scale and nature of intervention necessary to raise the level of Indigenous economic participation in the face of growing numbers moving into the working-age group? In short, what does the future Pilbara labour market look like, and where, in terms of numbers and composition, are Indigenous workers likely to fit?

There is a further question that is more universal in scope than the specifics of labour force participation, but it nonetheless arises out of the poor employment status that has been revealed because of the links between this, low incomes, and consequent high welfare dependency. This issue concerns the cost to government, and to people themselves, if social and economic conditions remain the same as currently experienced. Basically, the impost in terms of providing income support and other welfare payments, as well as program support in areas of health, housing and CDEP in particular, and the endless churning through the criminal justice system will simply escalate in line with the growth in population. On the other hand, if Indigenous people had more jobs at higher occupational levels, then they would be able to meet many of the basic needs that governments now provide for, from their own incomes, with the added bonus that many of the more negative fiscal expenditures would diminish.

Some estimate of this opportunity cost to government of simply continuing business as usual is provided here in the form of welfare dependency rates and associated estimates of dollar amounts. What is not costed though, is the potentially greater public impost of excess disease burden, infrastructure replacement, and foregone educational outcomes due to the continued and growing marginalisation of Indigenous people within the regional economy. It is important to recognise that the policy options for addressing this situation are not cost neutral – expenditure will grow either in response to declining economic status, or in order to enhance it. Whatever the case, a fiscal response is unavoidable.

An essential component of the drive to open up areas of the regional labour market to Indigenous employment is the need to tackle much deeper structural hurdles if Indigenous people are to successfully compete for skilled mainstream jobs with other residents (and potential in-migrants, both Indigenous and non-Indigenous). These include poor literacy and numeracy levels, which in part reflect low school participation and attendance levels. Also for noting is the fact of continuing high adult morbidity and mortality – if a 15-year-old Indigenous male in the region has only a 50 per cent chance of reaching retirement age, then the physical limitations on prolonged and full participation in the workforce become all too apparent, especially if we add to this the high rates of morbidity and disability that are prevalent throughout the prime working ages. One very practical implication of this premature mortality is a reduction in lifetime earning capacity, including the accumulation of superannuation. This diminishes the ability to accumulate assets and reduces the flow of intergenerational wealth, thereby perpetuating poverty traps.

Of course, not all regional aspirations point in the direction of mainstream workforce participation. With growing access to traditional lands, many people are making lifestyle choices and placing their emphasis on continuing ties to country, and the customary social and economic activities that stem from this. Where this intersects with mining activity, as in the case of heritage work, this may provide a source of meaningful engagement along with intermittent income. However, there is a need to explore other means of commercialising the customary sector in ways that require, rather than hamper, its sustainability. The arts industry and cultural education are obvious examples, but land management and work in the Indigenous organisation sector may provide more labour intensive and potentially widespread opportunities. Elsewhere in remote Australia, this combination of work in the private, public, and customary sectors has been referred to as a hybrid economy (Altman 2005). Whatever the case in the Pilbara, against the background of an expanding working-age population, the additional work opportunities generated by such activities should be seen as an essential component of the overall push to raise the level of Indigenous labour force participation. It is unlikely that the mining industry can achieve this alone.

The targets that have been set by Pilbara Iron and others in the region to enhance the level of Indigenous employment are difficult to translate into a whole-of-region estimate of labour demand due to the lack of a comprehensive database incorporating information from all key employers in the region. However, from the information provided by Pilbara Iron, and from what can be gleaned from BHP Billiton public documentation, it appears that these two largest employers combined require an additional 665 Indigenous workers over the next eight years in order to meet their targets. If achieved, this would double the present (2006) number of Indigenous people employed full-time in the mainstream Pilbara labour market. The supply-side questions that stem from this concern the limits to achievability, and the implications (even if achieved) for overall socioeconomic status.

Education and training

The polarisation of employment opportunity in the Pilbara between Indigenous and non-Indigenous people has many of its antecedents in relative educational status. While the historic reality is that many older Indigenous adults in the region have never attended school, it is equally true that many of those presently of compulsory school age do not attend school on a regular basis. Also apparent is a relative lack of progression through the school system to the crucial years of matriculation. While indicative data exist, one concern is that data on the practical outcomes of low school participation, as measured by literacy and numeracy achievement, are not publicaly available at the geographic scale of even the whole Pilbara, let alone that of its sub-components. Of course, the Western Australia Department of Education knows the details along with individual school boards and parents, but from a regional planning perspective where an attempt is being made to establish the overall quantum and composition of needs, as well as the interconnections between human capital variables and economic outcomes, this presents a significant gap in public knowledge.

Despite relatively low school participation, it remains the case that an estimated 2800 Indigenous adults in 2006 would have had some level of schooling through to Year 10 or above (although only 604 would have achieved Year 12). While the appropriate cross-tabulation has not been established, it seems reasonable to assume that the other estimate of almost 2100 Indigenous adults in work in 2006 (including CDEP) would be drawn from this 'educated' group, leaving some 700 individuals with Year 10 or above either unemployed or not in the labour force. It also means, of course, that an additional 1500 or so individuals also exist who have schooling levels below Year 10 and who therefore (with prevailing school attendance rates and literacy/numeracy achievement) present a sizeable remedial group if they are to be prepared for mainstream workforce participation.

Of course, age at leaving school, and even highest year of schooling completed, does not necessarily equate with grade level achievement. As the indicative WALNA data show, at best barely two-thirds of Indigenous Year 7 students in very remote parts of Western Australia achieve national benchmark levels in reading. On top of this, 286 Indigenous students in Year 8 to Year 10 regularly attend school, while only around one-third of these will continue on to Year 12 (95 students). If even a fraction of these are achieving at below standard levels, then this means that the numbers exiting the Pilbara education system with competencies at Year 12 level are almost certainly less than 100 each year. Against the estimated requirements for Pilbara Iron and BHP Billiton alone for an additional 665 Indigenous workers by 2013 to meet Indigenous employment targets, this suggests that the local flow of individuals with capacity to compete in the mainstream labour market is barely sufficient to match labour demand.

Somewhat similar calculations can be made in regard to VET sector output, although here the indications are more promising. While module load completion rates do not provide a direct measure of successful final outcomes in terms of producing qualified individuals, if the Indigenous rate observed for the mine hinterland is applied to Indigenous enrolments, then this suggests a potential future output of around 300 individuals emerging from the VET system, mostly at certificate levels I to III. While this would convert to an increase in the current estimate of 130 Indigenous adults in the region with post-secondary qualifications, many of these may well be the same people. Also, it is not known how many of those engaged in training already form part of the regional workforce, either with jobs in the mainstream or via CDEP.

Health status

Reference has been made already to the economic impacts of poor health status and high adult mortality. In the social epidemiology literature, this is a well-established self-reinforcing relationship (Berkman & Kawachi 2000; Marmot & Wilkinson 1999), and at least part of the project to enhance Indigenous participation in the economy of the Pilbara is a need to address the effects of low socioeconomic status on ill-health, as well as the other way around, especially in terms of ensuring an adequate start in life (Zubrick et al. 2004). Estimates generated here of the numbers likely to be excluded from regular (or even any) employment due to poor health point to a figure that could be approaching the size of the mainstream Indigenous workforce. There is imprecision here owing to data and time constraints, but given the potential enormity of this observation, more work needs to be done to establish the true scale of health impacts, much in the same way that Rowbottom et al. (2003) were able to derive a regional estimate of diabetics.

With reference to the life expectancy estimates for Indigenous people in the Pilbara – 52 and 55 years for males in the East and West Pilbara, and 60 and 63 years for females in the East and West Pilbara – the physical limitations on prolonged and full participation in the workforce become all too apparent. If we add to this the fact of relatively high Indigenous morbidity rates commencing in young adulthood and rising throughout the prime working ages, then a pattern emerges of severe physical constraints on the ability of many in the community to engage in meaningful and sustained economic activity. From a labour market perspective, it is likely that these negative effects of poor health status commence long before individuals are eligible to join the workforce, as suggested by relationships, long-established, between the poor health status of Indigenous people and below average school performance. There is also the likelihood of less direct impacts on workforce participation such as the prospect that many individuals do not seek work due to responsibilities in caring for sick relatives.

Among the issues underlying health status, this report emphasises the significance of ongoing backlogs in achieving adequate environmental health infrastructure, of the need for improved outcomes from education and training, of the difficulties of achieving better nutritional status in the population given the high cost of food and low incomes, and of the ongoing debilitating effects and social disruption caused by excessive alcohol consumption. This latter observation is underlined by the the 1994 NATSIS which revealed that 89 per cent of respondents in the South Hedland ATSIC region regarded alcohol as the main local health problem (the second highest rate in the country) (ABS 1996a: 64). All of these issues reflect on social and economic conditions in the region that are the focus of policy intervention. This notwithstanding, many reported Indigenous health outcomes in the Pilbara remain notably behind the rest of the state and undermine the capacity for participation in regional economic development.

Crime

One link between recidivism and the regional society and economy is the degree to which convictions and interaction with police, courts and prisons reduce individual chances of participating successfully in the regional economy. Criminologists have long been interested in the relationship between unemployment and crime, though with a focus particularly on examining the effect of unemployment on criminal behaviour (Chapman et al. 2002; Weatherburn 2002). In contrast, economists interested in the Indigenous labour market have considered the effect of a criminal conviction on an individual's employment prospects, with Hunter and Borland (1999) finding a strong negative impact of arrest. Certainly, in an industry as safety-conscious as mining, prior conviction and any ongoing substance misuse can be highly deleterious.

Accordingly, the summary statistics from police records, court records, and prison records for residents of the Pilbara presented here allow for some estimate of the population for whom contact with the police and a criminal conviction might represent a barrier, or at least a brake, on social and economic participation. Research on the factors underlying high arrest rates among Indigenous people and the effect of these on employment prospects indicates that if governments are concerned about Indigenous social and economic wellbeing then a priority should be to ensure that they stay out of the criminal justice system (Hunter 2001; Hunter & Borland 1999). Clearly, in the Pilbara, this has yet to be achieved as the statistics indicate high levels of recorded contact with police and subsequent conviction via the courts system.

Among the more telling facts are the following: the total number of unique Indigenous individuals arrested in a year (1047) is almost the same as the number in mainstream employment; virtually half of all Indigenous males aged between 15 and 34 years of age are arrested at least once each year; and around 313 Indigenous adults are subject to some form of detention or supervisory order at any one time. These represent quite substantial impacts on regional participation. However, if just three categories of offence were eliminated (traffic, public order, and offences against justice procedures, all of which are regulatory in some way), then cases brought before the criminal justice system would be halved.

Among the factors that contribute to high arrest rates among Indigenous people, high unemployment (or lack of meaningful work) and poor educational achievement have been identified as the most prominent (Hunter 2001), although the effects of drugs and alcohol and a breakdown in adherence to rules of customary law are also factors that emerge from the regional interviews and resonate with the findings of Pilbara-based research by the Law Reform Commission of Western Australia (Trees 2004). As we have seen, all of these pre-requisites for high arrest rates are prominent among Indigenous people in the Pilbara.

For example, from the hospital separations data it is apparent that excess use of alcohol is prevalent, so it is not surprising, as already noted, that 89 per cent of respondents to the NATSIS in the South Hedland ATSIC region identified alcohol as the main local health problem (ABS 1996a: 64). At the same time, high rates of injury reported in hospitalisation data are consistent with levels of assault reported to police, as is the fact that 72 per cent of NATSIS respondents considered family violence to be a major problem (ABS 1996a: 70). Such observations point to a cycle of social dysfunction at the family and community level that is reflected in the level of interaction with the criminal justice system and in the statements of many interviewees.

One line of argument suggests that by deliberately seeking incarceration via their actions, Indigenous youth are engaging in an alternative rite of passage to manhood (Biles 1983), although Ogilvie and Van Zyl (2001) view detention not as a rite of passage but rather as simply another venue for the construction of identity among marginalised and bored adolescents who are desperate for change to their routine. Whatever the case, there is no doubt that individual efforts to break into the regional labour market will be hampered by the lack of a steady and progressive acquisition of work skills and experience that are so necessary for successful engagement.

The future

Despite unprecedented labour demand in the Pilbara, the capacity of local Indigenous people to benefit from this remains substantially constrained by their limited human capital. Not that mining employment is the universally preferred option, with other avenues and priorities expressed for participation in the regional economy. The point here, though, is that wherever participation is sought via the mainstream labour market, then many in the Indigenous population will continue to experience structural disadvantage in the absence of substantially enhanced intervention to redress historic exclusion.

Of course, in pursuit of a social licence to operate, major corporates are already active in engaging Indigenous workers. But such is the depth of supply-side disadvantage, that a major challenge lies ahead if they are to meet stated targets (certainly in a collective sense) given that they will come close to exhausting the available supply of local employable labour without investing further in remedial training and possibly lowering the 'fitness for work' requirements. Even then, if current targets were to be achieved, the additional jobs created in fulfilling employment quotas would suffice only to keep pace with the growth in Indigenous working-age population. Thus, while much might be accomplished by the mining sector in the years ahead in terms of enhanced Indigenous engagement, little change might be discernable in overall regional economic status, with a large component of the population remaining detached from mainstream opportunities.

The constraints on participation implied by this scenario range across the spectrum of social and economic conditions. To indicate the scale of some of these that have been quantified using public access data, Table 9.1 provides estimates of labour force exclusion in 2006, bearing in mind that the adult population for that year is projected to be 4759. What this underlines is that the vast majority of Indigenous adults in the Pilbara do not have full schooling, or a qualification, around half of adults remain outside the labour force, many are hospitalised at any one time, others are subject to chronic conditions requiring strict management regimes, many again (especially young males) are arrested and incarcerated, and feeding into this adult realm are relatively low achievers from the education system. In any event, the potential for prolonged and productive workforce participation on the part of young people is severely curtailed by premature mortality.

Table 9.1. Summary select Indigenous indicators of the scale of labour force exclusion: Pilbara region, 2006

Population aged 15 +	4759
Has no post-school qualification	4200
Has less than Year 10 schooling	1500
Not in the labour force	2190
Hospitalised each year (all Indigenous persons)	2800
Has diabetes (25 years and over)	1020
Has a disability	1020
Arrested each year	1050
In custody/supervision at any one time	310
Achieving Year 7 benchmark literacy (current school attendees)	60%
15 year old males surviving to age 65	<50%

From a policy perspective, levels of economic exclusion on the scale indicated here raise questions about the adequacy of government resourcing to meet the backlog of disadvantage that has so obviously accumulated in the Pilbara region. Analysis elsewhere in remote Australia has found this to be substantially wanting in important areas of capacity building such as education (Taylor & Stanley 2005). From a grass roots perspective, they raise questions about how Pilbara Indigenous peoples view their future prospects, and what factors they perceive to be contributory. Though inevitably partial at best, final comments from two members of the Pilbara Indigenous community provide some insight.

Interview segment 59

The future really comes back to what we are going to do now to fix the foundations up. If we are going to let it go as it is, and keep taking all the power away from parents, keep taking the power, and they need to restore our law, and they put that spirit and head back into the people and then I think you'll see a whole lot different ... start to give em power back so they can contribute to their mob, their tribe, to try and curb the way everything is going at the moment. So on two sides, on the local side here, I'm trying to restore, in the community, all the foundations, fix all that so we can have a good foundation and grow again. The way that history has taken that foundation out from under us, that tree is dying, and it will keep dying if we let it keep going. It's about us now altogether, government. Government is the main one, I can do all I can here, but if government, if it's not coming from the top then I am fighting a losing war. And that has to be not just in one community it has to be across the board, giving the authority back where it should be.

We need to go back to that 200-year business, going back and sitting on equal terms, and then we've got a bright future, all of us because we've dealt back here and fixed this foundation. If we don't then we haven't got a good future because its just going to carry on the way it is and things will get worse. Where is our economy going then? My people might just go bad. You hear talk of it, people sit around drinking and stuff, they are talking them sort of things. Why don't we go and do this, they all ripping our country up and we getting nothing back from it, just chicken feed. It's society now, the metropolitan areas and even the Pilbara are getting bigger, and what's happening to our community here, and where is all the ore coming from, here. Everything, all the mining company, what are they giving back to the people? We are missing out on the luxury that is coming from our land, and other people are enjoying it.

Interview segment 60

The economy is growing here and there is so much that could be done, so much. I've always found that the Pilbara has always been just mining and Aboriginal. If you take out all the local businesses and local town that doesn't deal with mining you are buggered, its just mining exploration, and it's the economy for the Australian people, but you can't get away from the fact of the traditional Indigenous owners, so you will always find Indigenous and mine people hand in hand, and that's part of our lifestyle since early 1960s whenever it started. The history of mining Indigenous relationships is sad, less access to land, less access to jobs, and we don't have the skills, the skills from the education and the school area, they're the major problems. I'll be frank, I'm in my forties, and I look at what's behind of me. I was taken away and I went to school. But education here is at a low level, and its nowhere near what our future has to have here, and that's mining rights and knowing about the land, and what you have to do with the mining, and the government. Unfortunately the education failed us, it did, it failed us big time. So we have to continue on living until we are 60 years of age to continue working because our generation doesn't have that skill. And that's where we need to get that skill upgraded for our future.

We have elders and that age bracket of people who are 40 upwards who have the knowledge and education of working with mining. They are the ones that are speaking for the people and their children, instead of us being out there and talking to them and training them so they can understand what's expected of them when we die. We are doing it in our tradition and culture and ceremonies, but not, in what I call it, the Western world. But what of the education system of working our people, in knowing what you are about, what do you do, what do they do, what's their input into it. We know it's all survey and heritage, why haven't we got Aboriginal anthropologists, why haven't we got any Aboriginal person sitting up from our country in that mining company, being our liaison officer? Why isn't that so? You see what I am saying? It's not being handed down to the next one, it's not happening, I don't see it happening.

References

Aboriginal Training and Liaison (ATAL) 2005. Indigenous Employment Strategy, ATAL, Dampier, WA.

Altman, J. C. 2001. 'Economic development of the Indigenous economy and the potential leverage of native title', in B. Keon-Cohen (ed.), Native Title in the New Millennium, Native Title Research Unit, Australian Institute of Aboriginal and Torres Strait Islander Studies, Canberra.

—— 2005. 'Development options on Aboriginal land: sustainable Indigenous hybrid economies in the twenty-first century', in L. Taylor, G. K. Ward, G. Henderson, R. Davis and L.A. Wallis (eds), The Power of Knowledge: The Resonance of Tradition, Aboriginal Studies Press, Canberra.

Argus Research 2004. Western Australian Development Projects: Employment Demand and Predicted Skill Requirements 2003–2007, A Report prepared by Argus Research for the Western Australian Department of Education and Training, Perth.

Australian Bureau of Statistics (ABS) 1996a. 1994 National Aboriginal and Torres Strait Islander Survey South Hedland Region ATSIC Region, Cat. no. 4196.0.00.025, ABS, Canberra.

—— 1996b. 1994 National Aboriginal and Torres Strait Islander Survey Warburton Region ATSIC Region, Cat. no. 4196.0.00.023, ABS, Canberra.

—— 1998. Disability, Ageing and Carers Summary Tables, Western Australia 1998, Cat. no. 4430.0, ABS, Canberra.

—— 2002a. Housing and Infrastructure in Aboriginal and Torres Strait Islander Communities 1999, Australia, Cat. no. 4710.0, ABS, Canberra.

—— 2002b. Western Australian Indigenous Profiles, 2001, Cat. no. 2002.0, ABS, Canberra.

—— 2003. Criminal Courts, Cat. no. 4513.0, ABS, Canberra.

—— 2004. Deaths 2003, Cat. no. 3302.0, ABS, Canberra.

ABS and Australian Institute of Health and Welfare (AIHW) 2003. The Health and Welfare of Australia's Aboriginal and Torres Strait Islander Peoples 2003, Cat. no. 4704.0, ABS, Canberra.

ABS and Centre for Aboriginal Economic Policy Research (CAEPR) 1996. 1994 National Aboriginal and Torres Strait Islander Survey: Employment Outcomes for Indigenous Australians, Cat. no. 4199.0, ABS, Canberra.

Bell, M. 1992. Demographic Projections and Forecasts In Australia: A Directory and Digest, Australian Government Publishing Service, Canberra.

Bell, M. 2001. 'Understanding circulation in Australia', Journal of Population Research, 18 (1): 1–19.

Bell, M. and Maher, C. 1995. Internal Migration in Australia 1986–1991: The Labour Force, Australian Government Publishing Service, Canberra.

Bell, M. and Ward, G. 2000. 'Comparing temporary mobility with permanent migration', Tourism Geographies, 2 (1): 87–107.

Berkman, L. F. and Kawachi, I. 2000. Social Epidemiology, Oxford University Press, New York.

BHP Billiton Iron Ore 2003. Working Towards Sustainability in Western Australia: BHP Billiton and Boodarie Iron Health and Safety Environment and Community Report 2003, BHP Billiton External Affairs, Perth.

Biles, D. 1983. Groote Eylandt Prisoners: A Research Report, Australian Institute of Criminology, Canberra.

Birckhead, J. 1999. 'Brief encounters: doing rapid ethnography in Aboriginal Australia', in S. Toussaint and J. Taylor (eds), Applied Anthropology in Australasia, University of Western Australia Press, Nedlands, WA.

Clements, K. W. and Johnson, P. L. 1999. 'Minerals and regional employment in Western Australia', Economic Research Centre Discussion Paper No. 99/19, Department of Economics, University of Western Australia, Perth.

Chapman, B., Weatherburn, D., Kapuscinski, C. A., Chilvers, M. and Roussel, S. 2002. 'Unemployment duration, schooling and property crime', CEPR Discussion Paper No. 447, Centre for Economic Policy Research, ANU, Canberra.

Commonwealth of Australia 1999. The National Indigenous Housing Guide: Improving the Living Environment for Safety, Health and Sustainability, Commonwealth, State and Territory Housing Ministers' Working Group on Indigenous Housing, Canberra.

Coombs, H. C., McCann, H., Ross, H. and Williams, N. M. (eds) 1989. Land of Promises: Aborigines and Development in the East Kimberley, Aboriginal Studies Press, Canberra.

Cousins, D. and Nieuwenhuysen, J. 1984. Aboriginals and the Mining Industry: Case Studies of the Australian Experience, George Allen & Unwin, Sydney.

Daly, A. E. 1995. Indigenous and Torres Strait Islander People in the Australian Labour Market 1986 and 1991, Cat. no. 6253.0, ABS, Canberra.

Dillon, M. 1990. 'Social impact at Argyle: genesis of a public policy', in R. A. Dixon and M. C. Dillon (eds), Aborigines and Diamond Mining: The

Politics of Resource Development in the East Kimberley Western Australia, University of Western Australia Press, Nedlands, WA.

Disability Services Commission 1998. Profile of Disability for 1998: Pilbara Region, Disability Services Commission, Perth.

Divarakan-Brown, C. 1985. 'Premature ageing in the Aboriginal community', Proceedings of the Annual Conference of the Australian Association of Gerontology, 20: 33–4.

Earle, R. and Earle, L. D. 1999. 'Male Indigenous and non-Indigenous ageing: a new millennium community development challenge', South Pacific Journal of Psychology, 11 (2): 13–23.

Edmunds, M. 1989. They Get Heaps: A Study of Attitudes In Roebourne Western Australia, Aboriginal Studies Press, Canberra.

Fernandez, J. A. 2003. Regional Chart Supplement to Aboriginal Involvement In the Western Australian Criminal Justice System 2001, Crime Research Centre, University of Western Australia, Crawley, WA.

Fernandez, J. A., Ferrante, A. M., Loh, N. S. N., Maller, M. G. and Valuri, G. M. 2003. Crime and Justice Statistics for Western Australia: 2003, Crime Research Centre, University of Western Australia, Crawley, WA.

Fernandez, J. A. and Loh, N. S. N. 2001. Crime and Justice Statistics for Western Australia: 2001, Crime Research Centre, University of Western Australia, Crawley, WA.

George, K. 2003. 'Exclusive rights: ongoing exclusion in resource rich Aboriginal Australia', Australian Psychiatry, 11 (supplement): 9–12.

Government of Western Australia 2005. Prospect Magazine, March–May 2005, Department of Industry and Resources, Perth.

Gracey, M. and Spargo, R. M. 1987. 'The state of health of Aborigines in the Kimberley region', Medical Journal of Australia, 146: 200–4.

Graham (Polly) Farmer Foundation 2004. Annual Report 2004, Graham (Polly) Farmer Foundation Inc., Cottesloe, WA.

Hames Sharley 2004. Pilbara Workforce Delivery Strategy Technical Paper Two, Hames Sharley, Perth.

Harvey, B. 2001. 'Economic options in mining development negotiations II', in B. Keon-Cohen (ed.), Native Title in the New Millennium, Native Title Research Unit, Australian Institute of Aboriginal and Torres Strait Islander Studies, Canberra.

Harvey, B. 2002. 'New competencies in mining: Rio Tinto's experience', paper presented at the Council of Mining and Metallurgical Congress, 27–28 May 2002, Cairns.

Harvey, B. and Brereton, D. 2005. 'Emerging models of community engagement in the Australian minerals industry', Paper presented at the International Conference on Engaging Communities, August 2005, Brisbane.

Holcombe, S. 2004. 'Early Indigenous engagement with mining in the Pilbara: lessons from a historical perspective', CAEPR Working Paper No. 24, CAEPR, ANU, Canberra.

Hunter, B. H. 1996. 'The determinants of Indigenous employment outcomes: the importance of education and training', CAEPR Discussion Paper No. 115, CAEPR, ANU, Canberra.

—— 2001. Factors Underlying Indigenous Arrest Rates, NSW Bureau of Crime Statistics and Research, Attorney General's Department, Sydney.

—— 2004. Indigenous Australians in the Contemporary Labour Market 2001, Cat. no. 2052.0, ABS, Canberra.

Hunter, B. H. and Borland, J. 1999. 'The effect of arrest on Indigenous employment prospects', Crime and Justice Bulletin No. 45, NSW Bureau of Crime Statistics and Research, Attorney General's Department, Sydney.

Hunter, B. H. and Daly, A. E. 1998. 'Labour market incentives among Indigenous Australians: the cost of job loss versus the gains from employment', CAEPR Discussion Paper No. 159, CAEPR, ANU, Canberra.

Hunter, B. H. and Schwab, R. G. 2003. 'Practical reconciliation and recent trends in Indigenous education', CAEPR Discussion Paper No. 249, CAEPR, ANU, Canberra.

Hunter, B. H. and Taylor, J. 2004. 'Indigenous employment forecasts: implications for reconciliation', Agenda, 11 (2): 179–92.

Jones, R. 1994. The Housing Need of Indigenous Australians, 1991, CAEPR Research Monograph No. 8, CAEPR, ANU, Canberra.

Kakadu Region Social Impact Study (KRSIS) 1997. Report of the Indigenous Project Committee, Office of the Supervising Scientist, Canberra.

Kesteven, S. 1986. 'The project to monitor the social impact of uranium mining on Aboriginal communities in the Northern Territory', Australian Aboriginal Studies, 1986 (1): 43–5.

Kinfu, Y. and Taylor, J. 2002. "Estimating the components of Indigenous population change, 1996–2001', CAEPR Discussion Paper No. 243, CAEPR, ANU, Canberra.

Kinfu, Y. and Taylor, J. 2005. 'On the components of Indigenous population change', Australian Geographer, 36 (2): 233–55.

Linge, G. J. R. 1980. 'From vision to pipe dream: yet another northern miss', in J. N. Jennings and G. J. R. Linge (eds), Of Time and Place: Essays In Honour of O. H. K. Spate, ANU Press, Canberra.

Loh, N. and Ferrante, A. 2001. Indigenous Involvement in the Western Australia Criminal Justice System: A Statistical Review, 2000, Report for the Indigenous Justice Council, Crime Research Centre, University Of Western Australia, Crawley, WA.

Marmot, M. and Wilkinson, R. G. 1999. Social Determinants of Health, Oxford University Press, New York.

Martin, D. and Taylor, J. 1996. 'Ethnographic perspectives on the enumeration of Indigenous people in remote Australia', Journal of the Australian Population Association, 13 (1): 17–33.

McCorry, M. 2004. Success Stories: An Independent Assessment of The Partnerships For Success, Report to the Graham (Polly) Farmer Foundation, Cottesloe, WA.

Morphy, F. 2002. 'When systems collide: the 2001 Census at a Northern Territory outstation', in D. Martin, F. Morphy, W. Sanders and J. Taylor, Making Sense of the Census: Observations of the 2001 Enumeration in Remote Aboriginal Australia, CAEPR Research Monograph No. 22, ANU E Press, Canberra.

Musk, A. W., de Klerk, N. H., Eccles, J. L., Hansen, J., and Shilkin, K. B. 1995. 'Malignant mesothelioma in Pilbara Aborigines', Australian Journal of Public Health, 19 (5): 520–2.

National Centre for Social Applications of GIS 2003. Indigenous Housing Need: Homelessness, Overcrowding and Affordability: 2001 Census Analysis, Final Report to ATSIC, National Centre for Social Applications of GIS, University of Adelaide, Adelaide.

Ogilvie, E. and Van Zyl, A. 2001. 'Young Indigenous males, custody and the rites of passage', Trends and Issues in Crime and Criminal Justice No. 204, Australian Institute of Criminology, Canberra.

Olive, N. 1997. Karijini Mirlilirli: Aboriginal Histories from the Pilbara, Fremantle Arts Centre Press, Fremantle.

Parriman, F. and Daley, D. 1999. 'Indigenous Community Supervision Agreements in Western Australia', Paper presented at the Best Practice Interventions in Corrections for Indigenous People, a conference convened by the Australian Institute of Criminology and the Department of Correctional Services SA, 13–15 October 1999, Adelaide.

Phibbs, P. 1989. 'Demographic-economic impact forecasting in non-metropolitan regions: an Australian example', in P. Congdon and P. Batey (eds),

Advances in Regional Demography: Information, Forecasts, Models, Belhaven Press, London.

Pholeros, P., Rainow, S. and Torzillo, P. 1993. Housing for Health: Towards a Healthy Living Environment for Indigenous Australia, Health Habitat, Newport Beach, NSW.

Pilbara Development Commission 2004. Housing and Land Snapshot Update March 2004, Pilbara Development Commission, Port Hedland, WA.

Pilbara Population Health Unit 2004. Pilbara Health Profile, The Pilbara Population Health Unit, South Hedland, WA.

Ross, H. 1990. 'Progress and prospects in Indigenous social impact assessment', Australian Aboriginal Studies, 1990 (1): 11–17.

Rowbottom, J., Coffin, J., Dwyer, K., Larson, A. and Pain, V. 2003. Get em Waba: Pilbara Chronic Disease Operational Plan 2003, Combined Universities Centre for Rural Health and Pilbara Division of General Practice, WA.

Rowse, T. 2002. Indigenous Futures: Choice and Development for Aboriginal and Islander Australia, University of New South Wales Press Press, Sydney.

Saggers, S. and Gray, D. 2001. Port Hedland and Roebourne Substance Misuse Services Review, National Drug Research Institute, Curtin University of Technology, Perth.

Sanders, W. 2002. 'Adapting to circumstance: the 2001 census in the Alice Springs town camps', in D. Martin, F. Morphy, W. Sanders and J. Taylor, Making Sense of the Census: Observations of the 2001 Enumeration in Remote Aboriginal Australia, CAEPR Research Monograph No. 22, ANU E Press, Canberra.

Schwab, R. G. 1998. 'Educational "failure" and educational "success" in an Indigenous community', CAEPR Discussion Paper No. 161, CAEPR, ANU, Canberra.

Steering Committee for the Review of Government Service Provision (SCRGSP) 2005. Report on Government Services 2005, Indigenous Compendium, Productivity Commission, Canberra.

Smith, D. E. 1991. 'Toward an Indigenous household expenditure survey: conceptual, methodological and cultural considerations', CAEPR Discussion Paper No. 10, CAEPR, ANU, Canberra.

Storey, K. 2001. 'Fly-in/fly-out and fly-over: mining and regional development in Western Australia', Australian Geographer, 32 (2): 133–48.

Taylor, J. 1997. 'The relative economic status of Indigenous people in Western Australia, 1991–96', CAEPR Discussion Paper No. 157, CAEPR, ANU, Canberra.

—— 1998. 'Measuring short-term population mobility among indigenous Australians: options and implications', Australian Geographer, 29 (1): 125–37.

—— 1999. 'Aboriginal people in the Kakadu region: social indicators for impact assessment', CAEPR Working Paper No. 4, CAEPR, ANU, Canberra.

—— 2004a. Aboriginal Population Profiles for Development Planning in the Northern East Kimberley, CAEPR Research Monograph No. 23, ANU E Press, Canberra.

—— 2005a. 'Indigenous labour supply and regional industry', in D. Austin-Broos and G. Macdonald (eds), Culture and Economy in Aboriginal Australia, Sydney University Press, Sydney

—— 2004b. Social Indicators for Aboriginal Governance: Insights from the Thamarrurr Region, Northern Territory, CAEPR Research Monograph No. 24, ANU E Press, Canberra.

—— 2005b. 'When systems collide: the enumeration of Indigenous people in remote Australia', Plenary address to the International Association of Official Statistics Conference on Measuring Small and Indigenous Populations, April 2005, Wellington, NZ.

Taylor, J. and Bell, M. 1999. 'Changing places: Indigenous population movement in the 1990s', CAEPR Discussion Paper No. 189, CAEPR, ANU, Canberra.

—— and —— 2001. 'Towards a composite estimate of Cape York's Indigenous population', QCPR Discussion Paper No. 1, Queensland Centre for Population Research, University of Queensland, St Lucia.

—— and —— 2003. 'Options for benchmarking ABS population estimates for Indigenous communities in Queensland', CAEPR Discussion Paper No. 243, CAEPR, ANU, Canberra.

—— and —— 2004. 'Continuity and change in Indigenous Australian population mobility', in J. Taylor and M. Bell (eds), Population Mobility and Indigenous Peoples in Australasia and North America, Routledge, London.

Taylor, J. Bern, J. and Senior, K. 2000. Ngukurr at the Millennium: A Baseline Profile for Social Impact Planning in South-East Arnhem Land, CAEPR Research Monograph No. 18, CAEPR, ANU, Canberra.

Taylor, J. and Hunter, B. 1998. The Job Still Ahead: Economic Costs of Continuing Indigenous Employment Disparity, ATSIC, Canberra.

Taylor, J. and Roach, L. 1994. 'The relative economic status of Indigenous people in Western Australia, 1986–91', CAEPR Discussion Paper No. 59, CAEPR, ANU, Canberra.

Taylor, J. and Stanley, O. 2005. 'The opportunity costs of the status quo in the Thamarrurr Region', CAEPR Working Paper No. 28, CAEPR, ANU, Canberra.

Taylor, J. and Westbury, N. 2000. Indigenous Nutrition and the Nyirranggulung Health Strategy in Jawoyn Country, CAEPR Research Monograph No. 19, CAEPR, ANU, Canberra.

Treadgold, M. L. 1988. 'Intercensal change in Aboriginal incomes', Australian Bulletin of Labour, 14 (4): 592–609.

Trebeck, K. 2003. 'Corporate social responsibility, Indigenous Australians and mining', The Australian Chief Executive, Committee for Economic Development of Australia, Melbourne.

Trees, K. 2004. 'Contemporary issues facing customary law and the general legal system: Roebourne – a case study', Background Paper No. 6, Law Reform Commission of Western Australia, Perth.

Unwin, E., Codde, J., Swensen, G. and Saunders, P. 1997. Alcohol-Caused Deaths and Hospitalisation in Western Australia by Health Services, Health Department WA and WA Drug Abuse Strategy Office, Perth.

Walsh, F. and Mitchell, P. 2002. Planning for Country: Cross-Cultural Approaches to Decision-Making on Aboriginal Lands, IAD Press, Alice Springs.

Wangka Maya Pilbara Aboriginal Language Centre (WMPALC) 2001. Wumun Turi: Pilbara Aboriginal Women's Stories, WMPALC, South Hedland, WA.

Watson, J. Ejueyitsi, V. B. and Codde, J. P. 2001. A Comparative Overview of Indigenous Health in Western Australia, Epidemiology Occasional Paper No. 15, Department of Health, Perth.

Weatherburn, D. 2002. 'The impact of unemployment on crime', in P. Saunders and R. Taylor (eds), The Price of Prosperity: The Economic and Social Costs of Unemployment, University of New South Wales Press Press, Sydney.

Western Australia Department of Health 2005. Overview of the Major Causes of Mortality for Pilbara Public Health Unit Area Aboriginal Residents, Epidemiology Branch, HIC, Department of Health, Perth.

Western Australia Department of Industry and Resources 2004. Western Australian Mineral and Petroleum Statistics Digest 2003–04, Department of Industry and Resources, Perth.

Western Australia Department of Justice 2002. Adult Offender Annual Statistical Report for Financial Year 2001/2002, Department of Justice, Perth.

Western Australian Planning Commission 2005. Western Australia Tomorrow: Population projections for Planning Regions 2004 to 2031 and Local Government Areas 2004 to 2021, Western Australian Planning Commission, Perth

Willis, J. 1995. 'Fatal attraction: do high technology treatments for end stage renal disease benefit Aboriginal people in central Australia?' Australian and New Zealand Journal of Public Health, 19 (6): 603–9.

Wilson, J. 1980. 'The Pilbara Aboriginal Social Movement: an outline of its background and significance', in R. Berndt and C. Berndt (eds), Aborigines of the West: Their Past and Present, University of Western Australia Press, Perth.

Zubrick, S. R., Lawrence, D. M., Silburn, S. R., Blair, E., Milroy, H., Wilkes, T., Eades, S., D'Antoine, H., Read, A., Ishiguchi, P. and Doyle, S. 2004. The Western Australian Aboriginal Child Health Survey: The Health of Aboriginal Children and Young People, Telethon Institute for Child Health Research, Perth.

CAEPR Research Monograph Series

1. *Aborigines in the Economy: A Select Annotated Bibliography of Policy-Relevant Research 1985–90*, L. M. Allen, J. C. Altman, and E. Owen (with assistance from W. S. Arthur), 1991.

2. *Aboriginal Employment Equity by the Year 2000*, J. C. Altman (ed.), published for the Academy of Social Sciences in Australia, 1991.

3. *A National Survey of Indigenous Australians: Options and Implications*, J. C. Altman (ed.), 1992.

4. *Indigenous Australians in the Economy: Abstracts of Research, 1991–92*, L. M. Roach and K. A. Probst, 1993.

5. *The Relative Economic Status of Indigenous Australians, 1986–91*, J. Taylor, 1993.

6. *Regional Change in the Economic Status of Indigenous Australians, 1986–91*, J. Taylor, 1993.

7. *Mabo and Native Title: Origins and Institutional Implications*, W. Sanders (ed.), 1994.

8. *The Housing Need of Indigenous Australians, 1991*, R. Jones, 1994.

9. *Indigenous Australians in the Economy: Abstracts of Research, 1993–94*, L. M. Roach and H. J. Bek, 1995.

10. *The Native Title Era: Emerging Issues for Research, Policy, and Practice*, J. Finlayson and D. E. Smith (eds), 1995.

11. *The 1994 National Aboriginal and Torres Strait Islander Survey: Findings and Future Prospects*, J. C. Altman and J. Taylor (eds), 1996.

12. *Fighting Over Country: Anthropological Perspectives*, D. E. Smith and J. Finlayson (eds), 1997.

13. *Connections in Native Title: Genealogies, Kinship, and Groups*, J. D. Finlayson, B. Rigsby, and H. J. Bek (eds), 1999.

14. *Land Rights at Risk? Evaluations of the Reeves Report*, J. C. Altman, F. Morphy, and T. Rowse (eds), 1999.

15. *Unemployment Payments, the Activity Test, and Indigenous Australians: Understanding Breach Rates*, W. Sanders, 1999.

16. *Why Only One in Three? The Complex Reasons for Low Indigenous School Retention*, R. G. Schwab, 1999.

17. *Indigenous Families and the Welfare System: Two Community Case Studies*, D. E. Smith (ed.), 2000.

18. *Ngukurr at the Millennium: A Baseline Profile for Social Impact Planning in South-East Arnhem Land*, J. Taylor, J. Bern, and K. A. Senior, 2000.

19. *Aboriginal Nutrition and the Nyirranggulung Health Strategy in Jawoyn Country*, J. Taylor and N. Westbury, 2000.

20. *The Indigenous Welfare Economy and the CDEP Scheme*, F. Morphy and W. Sanders (eds), 2001.

21. *Health Expenditure, Income and Health Status among Indigenous and Other Australians*, M. C. Gray, B. H. Hunter, and J. Taylor, 2002.

22. *Making Sense of the Census:Observations of the 2001 Enumeration in Remote Aboriginal Australia*, D. F. Martin, F. Morphy, W. G. Sanders and J. Taylor, 2002.

23. *Aboriginal Population Profiles for Development Planning in the Northern East Kimberley* J. Taylor, 2003.

24. *Social Indicators for Aboriginal Governance: Insights from the Thamarrurr Region, Northern Territory*, J. Taylor, 2004.

For information on CAEPR Discussion Papers, Working Papers and Research Monographs (Nos 1-19) please contact:

Publication Sales, Centre for Aboriginal Economic Policy Research,
The Australian National University, Canberra, ACT, 0200

Telephone: 02–6125 8211
Facsimile: 02–6125 2789

Information on CAEPR abstracts and summaries of all CAEPR print publications and those published electronically can be found at the following WWW address:

http://www.anu.edu.au/caepr/

www.ingramcontent.com/pod-product-compliance
Lightning Source LLC
Chambersburg PA
CBHW061238270326
41928CB00033B/3440